Lecture Notes in Statistics

Edited by S. Fienberg, J. Gani, J. Kiefer, and K. Krickeberg

3

Bruce D. Spencer

Benefit-Cost Analysis of Data Used to Allocate Funds

Springer-Verlag
New York Heidelberg Berlin

Lecture Notes in Statistics

Vol. 1: R. A. Fisher: An Appreciation. Edited by S. E. Fienberg and D. V. Hinkley. xi, 208 pages, 1980.

Vol. 2: Mathematical Statistics and Probability Theory. Proceedings 1978. Edited by W. Klonecki, A. Kozek, and J. Rosiński. xxiv, 373 pages, 1980.

Vol. 3: B. D. Spencer, Benefit-Cost Analysis of Data Used to Allocate Funds. viii, 296 pages, 1980.

Springer Series in Statistics

L. A. Goodman and W. H. Kruskal, Measures of Association for Cross Classifications. x, 146 pages, 1979.

J. O. Berger, Statistical Decision Theory: Foundations, Concepts, and Methods. vix, 420 pages, 1980.

Lecture Notes in Statistics

Edited by S. Fienberg, J. Gani, J. Kiefer, and K. Krickeberg

3

Bruce D. Spencer

Benefit–Cost Analysis of Data Used to Allocate Funds

Springer-Verlag
New York Heidelberg Berlin

Professor Bruce D. Spencer
Northwestern University
The School of Education
Evanston, Illinois 60201/USA

AMS Subject Classification: 62P25

Library of Congress Cataloging in Publication Data

Spencer, Bruce D
Benefit-cost analysis of data used to allocate funds.

(Lecture notes in statistics ; 3)
Based on the author's thesis, Yale.
Bibliography: p.
1. Revenue sharing—United States—Cost effectiveness. I. Title. II. Series.
HJ275.S59 336.1'85 80-19589

ISBN 0-387-**90511**-1 Springer-Verlag New York Heidelberg Berlin
ISBN 3-540-**90511**-1 Springer-Verlag Berlin Heidelberg New York

Printed in the United States of America.

9 8 7 6 5 4 3 2 1

Preface

This monograph treats the question of determining how much to spend for the collection and analysis of public data. This difficult problem for government statisticians and policy-makers is likely to become even more pressing in the near future. The approach taken here is to estimate and compare the benefits and costs of alternative data programs. Since data are used in many ways, the benefits are hard to measure. The strategy I have adopted focuses on use of data to determine fund allocations, particularly in the General Revenue Sharing program.

General Revenue Sharing is one of the largest allocation programs in the United States. That errors in population counts and other data cause sizable errors in allocation has been much publicized. Here we analyze whether the accuracy of the 1970 census of population and other data used by General Revenue Sharing should be improved. Of course it is too late to change the 1970 census program, but the method and techniques of analysis will apply to future data programs. In particular, benefit-cost analyses such as this are necessary for informed decisions about whether the expense of statistical programs is justified or not. For example, although a law authorizing a mid-decade census was enacted in 1976, there exists great doubt whether funds will be provided so a census can take place in 1985. (The President's Budget for 1981 allows no money for the mid-decade census, despite the Census Bureau's request for $1.9 million for planning purposes.) Is this decision in the national interest? Explicit benefit-costs analyses are severely needed.

Much of the present work is technical. The reader interested solely in policy implications is encouraged to read chapter 0 lightly and then proceed directly to chapter 7. For a thorough but non-technical course, I suggest chapter 0; chapter 1, first four sections at least; chapter 3 sections 1 and 2, then other sections as interest dictates; the first section of chapter 4; the first section or two of chapter 5; then chapters 6 and 7, skipping proofs all the while. The manuscript is heavily cross-referenced, so the reader should feel free to skip around, since work that is used later will be referenced at that time. In addition, guideposts placed throughout the work indicate parts that can be skipped on a first reading. An overview of the work is given in chapter 0.

Acknowledgments

This monograph has arisen from my doctoral dissertation for the Department of Statistics at Yale University. I wish to express my indebtedness to I. Richard Savage, who led me to the problem and provided continuing guidance and stimulation. I am grateful also to William Kruskal, Stephen Dresch, Francis Anscombe, and Thomas Jabine for reading the manuscript and for helpful comments.

This research was largely carried out while I was at Yale, where I was partly supported by university fellowships and by a National Science Foundation grant (SOC 75-15614). I wrote the final draft while at the Committee on National Statistics of the National Research Council - National Academy of Sciences and while at Northwestern University. I am grateful to colleagues of all three places for their encouragement and support.

Evanston, Illinois
March 1980

B. D. S.

TABLE OF CONTENTS

LIST OF TABLES

Chapter 0. Introduction

In 1970 the U.S. census cost more than $220 million and failed to count an estimated 5.3 million people. The 1980 census will cost a billion dollars and the size of the undercount, or the number of persons missed, is also certain to be in the millions. If the undercount were evenly distributed among geographic regions and population subgroups, matters would not be as serious as they are. In fact, the undercount is spread unevenly, with certain minority groups and geographic regions undercounted more than others. Census data are used in more than 100 formulas that allocate over $50 billion annually in federal aid. A greater than average undercount for a group or region can deny it its fair share of these funds. Since census data also determine apportionment of seats in the House of Representatives, undercount can cause a state to gain or lose a seat. National ethnic organizations worry that undercounts diminish their political influence by failing to indicate their true numbers.

Errors in census data clearly have significant impact, but how much should be spent to get good data? The 1970 census missed an estimated 2.5 percent of the population. Would it have been worthwhile to spend an extra $10 million to reduce the undercount? An extra $100 million?

Similar questions, not just about censuses but about other data collection activities as well, are relevant to statistical agencies in general, already hard pressed by increasing demands for a multitude of data. And the present conflict between demands and scarce resources can be expected to worsen as the demands for data outstrip the resources for obtaining and providing the data. The preferred way to cope with this problem is rational setting of statistical priorities (Moser 1977 ; National Research Council Costs and benefits of alternative data programs must be evaluated and compared.*

Because the data produced by statistical agencies form a social good and have many uses, the benefits of data quality are difficult to measure (Savage 1975). In setting statistical priorities it is necessary to measure the differences in the benefits arising from data packages of different quality (and cost). A reasonable way to proceed with the measurement is to focus on those areas where major benefits or losses could result from quality of data. Such a partial benefit analysis may not lead to precise conclusions but will aid in making better decisions. In particular, such analyses may help statistical agencies in supporting budget requests. Thus, reducing population undercount gives rise to many benefits, but the benefits from improved fund allocations alone may be substantial enough to justify the added cost.

The steps in benefit analysis of data are:

1. identify uses and users of data,
2. identify benefits arising from use of the data,
3. measure the benefits, and
4. analyze the sensitivity of the benefits to the quality (and cost) of the data.

The present work performs a benefit analysis for the 1970 census of population and also for several other programs producing data used for allocating funds. Allocation of funds by formula is a major and increasingly important use of statistical data. The State and Local Fiscal Assistance Act of 1972, commonly known as General Revenue Sharing (GRS), has used data to distribute over $55 billion to the 50 states, the District of Columbia, and some 39,000 local governments from 1972 to 1980.

---

* Other approaches, such as modern voting theory (Kramer and Klevorick 1973), may also be relevant but will not be pursued here.

The legislation requires that the allocations be determined by formulas using a variety of statistical data as input. But substantial inaccuracies in allocations arise from measurement errors, definition errors, and lack of timeliness in data. Interesting policy questions thus arise. For example, a recent National Research Council (1980) study of the postcensal estimates of population and per capita income for the 39,000 GRS areas questioned whether the estimates for certain classes of areas were so inaccurate that they should not be produced for use in GRS. The question we consider here is "should the data be improved?".

Our focus will be largely on the data used to allocate funds in GRS. To determine whether the data should be improved we first consider several major questions. The first question concerns how much it is worth to improve allocations. If X and Y are two sets of allocations but Y is more accurate, how much extra is it worth to obtain Y instead of X? To answer this question we adopt the perspective that Congressional intent would be realized if the allocations were based on error-free data.* The various conceptual problems addressed include: To whom is it worth spending more money? What are the benefits from improving the accuracy of the allocations? What are appropriate measures of the benefits? How can we compare the benefits of improved accuracy to costs of better data?

A second major question concerns the current accuracy of the data. Clearly, in deciding whether to improve the data it is important to know how good the data are to begin with. The present work draws upon a large body of literature to derive estimates of the biases, standard deviations, and correlations of the data series used to allocate funds in GRS.

A third question relates to the accuracy of the allocations. Since the allocations are determined exactly by the data and the accuracy of the data is known, the accuracy of the allocations can be determined. But the allocation formulas are incredibly

---

* Even if the allocations resulting from perfect data would not be optimal to Congress or others, those allocations play the roles of norms. Thus, in particular, evidence that an actual allocation is less than this norm is taken as evidence of an injustice or lack of equity. This same perspective is adopted in a recent report of the Office of Federal Statistical Policy and Standards (1978).

complex and several kinds of approximations will be required. We proceed in steps. First we develop the approximation theory, then we determine the structural relationship between accuracy in data and accuracy in allocations, then we estimate the accuracy in the allocations based on our estimates of data accuracy.

Once we have resolved these three major questions we can easily determine how much of the cost of improving data is justified by the improvement in allocations. We start with a relationship between data cost and accuracy. For example, spending D extra dollars on the census reduces undercount by U. We then calculate the increase in accuracy of allocations, and finally the benefit from the improvement in the allocations. The benefit is then compared to the cost to determine how much of the cost is justified by the benefit. In particular, we consider whether more money should have been spent to reduce undercount in the 1970 census. We also consider whether other data improvements would be advisable.

Other authors have performed benefit analyses for data, but the number of analyses is surprisingly small given the magnitude of the problem. The earliest identified analysis was performed by H. V. Muhsam (1956) for the uses of population forecasts in the planning of water supply facilities. Lave (1963) analyzes the benefits (increased profits) to the California raisin industry of accurate weather forecasts. The benefits arising from U. S. Department of Agriculture crop and livestock statistics are measured by Hayami and Peterson (1973) and Decanio (1978) from the viewpoint of crop inventory and production adjustment. Decanio is concerned with increased profits for the farmers while Hayami and Peterson use the Marshallian social welfare framework to measure <u>social</u> benefits. Bradford et al. (1974) is a more extensive analysis along the lines of Hayami and Peterson. Each of the above benefit analyses shared three features:

(i) Effects of errors in data were easy to model
(ii) Uses of data were sensitive to errors
(iii) Marketing mechanisms could be used to quantify the benefits.

Investigations of the effects of errors in data used to determine GRS allocations have been carried out by Robinson and Siegel (1979),

Siegel et al. (1977), Siegel (1975), Strauss and Harkins (1974), Hill and Steffes (1973), Stanford Research Institute (1974 a,b), and Savage and Windham (1974). In GRS the formulas used to determine the allocations are known explicitly so the consequences of known errors in data can be precisely calculated. However, quantification of the benefits from improving the accuracy of the allocations is extremely difficult. Essentially, an underallocation to one government is offset by an overallocation to another. A challenging aspect of benefit-cost analysis of data used for GRS allocations is lack of reliance on (iii) above.

This work extends in a number of directions the earlier analyses of data used in GRS. The scope is not limited to GRS, however, and the methodology and theory developed will apply to more general problems as well. Major features of the analysis are identified in the following outline.

Chapter 1: Loss Functions and Benefit Measurement

We seek to measure benefits from improved allocations in units comparable to the units of data cost (dollars). This entails various conceptual problems, which are analyzed. The benefits are perceived to take the form of increased social welfare and decreased inequity. We measure these benefits using loss functions to quantify the losses from inequity and decreased social welfare that arise from errors in allocations. Mathematical and statistical properties of the loss functions are studied. The loss function adopted measures the loss from inaccuracy in the allocations under a fixed-pie program (such as GRS, where the total allocation is fixed) by 1 percent of the sum of absolute deviations of actual allocations from error-free allocations. Interpretation and motivation of this loss function are provided.

Chapter 2: The Delta Method

The analytical problems encountered are complex and several kinds of approximations become necessary. The delta method, a traditional tool for such needs, is modified to apply to the complicated situations at hand. Applicability of the approximation methodology to the formulas for GRS is discussed.

Chapter 3: Data Used in GRS

To consider the effects of data quality upon GRS allocations it is of course imperative to study the quality of the data. Explicit stochastic models for errors (means, variances, and covariances) are obtained for the data used in GRS. Explicit error models are developed to allow separate consideration of individual sources of error, including census undercounts, response bias, response variance, sampling variance, and forecasting error. Recent estimates of state population undercoverage (Siegel et al. 1977) are incorporated.

Chapter 4,5: Interstate, Intrastate Allocations in GRS

The GRS allocations to states and to substate units are exactly determined by the input data. Using the delta method we develop explicit formulas expressing the accuracy of the allocations in terms of the accuracy of the data. That is, the moments (means and variances) of errors in allocation are expressed as functions of the moments of errors in the input data. This is the first time such analytical expressions have been derived. The methodology extends to other complex uses of data as wel

Chapter 6: Computations and Analyses

Using (i) the expressions relating the moments of errors in allocations to the moments of errors in data (chapters 4 and 5) and (ii) the numerical estimates of the moments of errors in data (chapter 3) we can routinely estimate the numerical values of the moments of errors in allocations. Means and variances of errors in allocation are estimated for all states and for a variety of local governments. The large number of local governments has necessitated restricting the substate analysis to just one state. The local governments considered here consist of all county governments in New Jersey and all municipalities in Essex County, N. J. Analysis for other counties would be straightforward, mechanically identical to Essex County.

Illustrative benefit-cost analyses are performed for several data programs. The analyses begin with a given relationship between data cost and accuracy. In particular, spending more or less money on data effects changes in the moments of the errors in data (chapter 3), hence changes in the moments of the errors in allocation (chapter 6). The loss function (chapter 1) can then be used to examine to what extent an increase (or decrease) in data cost was offset by the benefit from improving (or diminishing) the accuracy of the allocations. In this way we estimate the benefits for various data programs, including changing the levels of population undercount, reducing response errors in income reporting, improving the timeliness of population estimates, reducing the error in postcensal population and per capita income estimates, and reducing error in forecasts of tax collections. Several benefit-cost analyses are performed. For example, analysis shows that if certain coverage improvement programs had been used in the 1970 census, the benefits would have outweighed the costs.

Chapter 7: Policy Perspectives

Policy recommendations are made for the following issues: construction of allocation formulas, adjusting data for suspected biases (e.g., undercoverage), and improving data quality. Specific findings and conclusions include the following.

1. Priority should be given to minimizing the sum of absolute errors in allocation, while keeping individual errors within tolerance limits.
2. In tiered (including single-tiered) allocation programs data with uniform bias rates (in sign and size) should be used to determine allocations within any given tier. If analysis shows that one tier dominates in causing inequities, then data for that level should receive high priority for improvement. Analysis of GRS suggests that substate data are of secondary importance because they are dwarfed by errors in state data.
3. More evidence and analysis would be needed to support a decision to adjust the data underlying GRS for suspected biases. At this time, it cannot be recommended that population estimates be adjusted for estimated undercoverage.
4. Coverage improvement efforts for the 1980 census should be increased from the level of 1970 efforts.
5. Since conflicting or reinforcing aims of the legislators can effect a collinearity among the variables in an allocation formula, the formulas

should, where practicable, be constructed to exploit this effect to reduce the amount of data required.

Future research is also suggested.

*

The following notation is used throughout, except where noted. The symbol R^k denotes k-dimensional Euclidean space and vectors $(x_1 \ldots x_k)^T$ in R^k will be denoted by $\underset{\sim}{x}$ or on occasion as x, where the superscript ^{T} denotes matrix or vector transposition.

Summation will often be denoted by a ".", as

$$x. = \sum_i x_i \quad , \qquad (xy). = \sum_i x_i y_i \quad , \text{ and } \quad y_{i}. = \sum_j y_{ij} \quad .$$

The functions $(.)^+$ and $(.)^-$ are defined by

$$x^+ = \max\,(0,x) \quad \text{and} \quad x^- = \max\,(0,-x)$$

and the indicator function I_S over a set S is defined by

$$I_S(x) = \begin{cases} 1 & \text{if } x \text{ belongs to } S \\ 0 & \text{otherwise.} \end{cases}$$

For basic references on probability and statistical distribution theory the reader is referred to Feller (1968; 1971) and Rao (1973) respectively. The expectation of a real random variable X is denoted by EX, and the variance by Var X or V(X). If X has a normal distribution with mean zero and variance one its probability density function and cumulative distribution function will be represented by $\phi(.)$, $\Phi(.)$ respectively. If X and Y are random variables the covariance of X with Y is denoted by Cov(X,Y). The covariance matrix of a random vector $\underset{\sim}{Z}$ is denoted by $\text{Cov}(\underset{\sim}{Z})$.

Chapter 1 Loss Function and Benefit Measurement

§ 1.0 Introduction

The allocation of funds by formula is a major use of data collected by statistical agencies. As stated, the original General Revenue Sharing (GRS) program distributed more than $30 billion from 1972 to 1976 to 51 state governments (including the District of Columbia) and nearly 39,000 local governments. The allocations were made over time in six or twelve month "entitlement periods", with the total allocation in each entitlement period fixed (except for possible minor adjustments) by law. To determine the relative sizes of shares of the pie, GRS relied on formulas depending on values of data elements such as population, per capita income, and income tax collections. Appendix B describes the formulas in more detail.

In this chapter we consider ways of quantifying the benefits that arise from increasing the accuracy of the allocations. The purpose of quantifying the benefits is to enable the statistical agencies to compare the costs and benefits of alternative statistical programs. The analysis treats the heads of statistical agencies as a collective agent responding to the policy made by the Administration, by Congress, etc. This agent will be referred to as the Decision Maker. To permit comparison of costs and benefits of alternative statistical activities we will construct loss functions to measure the losses (negative benefits) that arise from errors in the data.

Allocation formulas will be denoted by $\underset{\sim}{f}, \underset{\sim}{g}, \underset{\sim}{h}, \ldots$ and are to be thought of as functions mapping data points $\underset{\sim}{Z}$ into n-demensional Euclidean space. The data point $\underset{\sim}{Z}$ is a long vector containing all values used to determine allocations to all recipients in a particular allocation period. By recipient we mean an entity which receives allocations; in GRS the recipients are state and local governments. The parameter n refers to the number of recipients and the ith coordinate of $\underset{\sim}{f}(\underset{\sim}{Z})$, $f_i(\underset{\sim}{Z})$, is the allocation to the ith recipient according to the formula $\underset{\sim}{f}$ and data $\underset{\sim}{Z}$. The vector of allocations intended by the policy-makers will be denoted by $\underset{\sim}{\theta}$ and called the vector of <u>optimal</u> allocations. For an allocation formula $\underset{\sim}{f}$, $M_f = \Sigma f_i$ $(M_\theta = \Sigma\theta_i)$ is the size of the allocated pie (optimal pie) and $\underset{\sim}{f} - \underset{\sim}{\theta}$ is the vector of <u>misallocations</u> to the recipients.

Analysis will begin from the assumption that the legislated formula is correct, in the sense that if $\underset{\sim}{Z}$ were observed without error then $\underset{\sim}{\theta}$ would be allocated by the formula. Usefulness of the analysis does not depend on this assumption. For if $\underset{\sim}{f}$ is the legislated formula, $\underset{\sim}{g}$ is the "correct" formula, $\underset{\sim}{Z}$ is the actual data and $\underset{\sim}{Z}^*$ is the error-free value of the data (so $\underset{\sim}{\theta} = \underset{\sim}{g}(\underset{\sim}{Z}^*)$) then we have

$$\underset{\sim}{f}(\underset{\sim}{Z}) - \underset{\sim}{\theta} = \underset{\sim}{f}(\underset{\sim}{Z}) - \underset{\sim}{f}(\underset{\sim}{Z}^*) + \underset{\sim}{f}(\underset{\sim}{Z}^*) - \underset{\sim}{g}(\underset{\sim}{Z}^*) .$$

Analysis of $\underset{\sim}{f}(\underset{\sim}{Z}) - \underset{\sim}{\theta}$ would proceed by consideration of $\underset{\sim}{f}(\underset{\sim}{Z}) - \underset{\sim}{f}(\underset{\sim}{Z}^*)$, $\underset{\sim}{f}(\underset{\sim}{Z}^*) - \underset{\sim}{g}(\underset{\sim}{Z}^*)$, and interactions between the two terms. The present analysis is a major part in such a procedure.

Since the Decision Maker will wish to compare losses arising from errors in allocation with costs of improving the data, the following properties are desirable:

P1. The loss function is scalar-valued.
P2. The loss function is measured on a scale such that the Decision Maker desires to minimize the cost of the data plus the expected loss from errors in allocation.
P3. The loss function is minimized when $\underset{\sim}{f} = \underset{\sim}{\theta}$.

Property P1 facilitates comparison of misallocation losses and data costs, since the latter are usually expressed in dollars. Dollars are also likely to be the most convenient medium for comparing benefits of different uses of data. Vector-valued loss functions have been studied (e.g. Keeney and Raiffa 1976) but the minimization described in P2 becomes difficult. In the special case that data collection costs can be attributed directly to the allocation program, property P1 is not essential. This is shown in the example of P. Redfern (1974) discussed in §1.1. The desirability of property P2 arises from the basic optimality principle of statistical decision theory: minimization of expected losses. Property P3 ensures that a formal analysis based on the loss function will be consistent with the goal of the allocation program: to allocation the amounts θ_i.

The purpose of this chapter is to determine a suitable loss function for application to the problem of deciding how much data quality is desirable when the data is used to determine fund allocations. First we consider the social issues and other authors' formulations of loss functions to measure losses from errors in allocation. Earlier constructions of loss functions for allocation processes have rested on the principles of either equity (§1.2) or social welfare (§1.1). Measuring increases in social welfare is generally difficult because the number of recipients is large, actual uses of the funds are hard to trace, and losses are not first order (overallocations balance underallocations). Measuring inequity is hard because the way policymakers (e.g. Congress) evaluate inequity is not understood.

Our approach will be to formulate different classes of possible loss functions and consider their respective merits. The considerations of merit will rest not only on social perceptions of loss but also on mathematical and statistical properties of the loss functions. In §1.3 several general classes of loss functions are proposed for measurement of losses from errors in allocations. One simple loss function measures loss proportional to the sum of the absolute errors in allocation. This loss function is motivated and interpreted (§1.3) and estimation of the proportionality constant is discussed (§1.4). Various mathematical and statistical properties of loss functions, including Fisher-consistency, are presented (§1.5). More general loss functions are considered and related to the absolute errors loss function (§1.6). Exponential loss functions are also discussed (§1.7). Sections 1.5 - 1.7 treat theoretical aspects of loss functions and provide additional justification for using the absolute errors loss function. These sections may be skipped on a first reading, as their results are not essential to later developments. The chapter concludes that the absolute errors loss function will be used in the sequel.

§ 1.1 Utility and Social Welfare

Fishburn (1968, p.335) writes that modern utility theory

> is concerned with people's choices and decisions... also with people's preferences and judgements of preferability, worth, value, goodness...
>
> The usual raw materials on which a utility theory... is based are an individul's preference-indifference relation $\lesssim$, read "is not preferred to", and a set X of elements $x,y,z,\ldots$ usually interpreted as decision alternatives or courses of action. $\lesssim$ is taken to be a binary relation on X, which simply says that if x and y are in X then exactly one of the following two statements is true:
>
> 1. $x \lesssim y$ (x is not preferred to y)
> 2. not $x \lesssim y$ (it is false that x is not preferred to y).
>
> ....A utility theory is essentially
>
> 1. a set of internally-consistent assumptions about X and the behavior of $\lesssim$ on X
> 2. the theorems that can be deduced from the assumptions.

Von Neumann and Morgenstern (1953, p.26) presented postulates implying the existence of a function $u: X \to R^1$ such that

(1.1) $\qquad x \lesssim y \quad$ if and only if $\quad u(x) \leq u(y)$

and

(1.2) $\qquad u(\alpha x + (1-\alpha)y) = \alpha u(x) + (1-\alpha)u(y) \qquad 0\leq\alpha\leq 1$

where $\alpha x + (1-\alpha)y$ is an element of X and represents the "randomized" event x with probability α and y with probability $1-\alpha$. A function u satisfying (1.1) is called a utility function.

The ranking on X induced by u is the same as the given preference-indifference relation. If a utility function also satisfies (1.2) it is called a von Neumann-Morgenstern utility function. In such a situation preference under uncertainty is reflected by expected utilities. Property P2 implies that from the perspective of the Decision Maker, the negative of $L(\underset{\sim}{f},\underset{\sim}{\theta})$ is a von Neumann-Morgenstern utility function. Note that if u satisfies (1.1) and (1.2) so does $v = \lambda u + \gamma$ where γ is any and λ is any positive constant. This implies that interpersonal comparisons of utility cannot be made by comparing the magnitudes of the individual utilities. For a lucid introduction see Luce and Raiffa (1957).

An interesting example of the social welfare-utility theory approach to the problem of constructing loss functions is that of Redfern (1974). A fixed pie M is divided among n local areas each with the same <u>actual</u> population size N. The allocation is <u>pro rata</u> to <u>estimated</u> population $\hat{N}_i$, so local area i receives an allocation of $M(\hat{N}_i)(\sum_j \hat{N}_j)^{-1}$. For each local area the welfare u from allocation y is

$$u(y) = K\,y^{1-B} \tag{1.3}$$

where K and B are constants, the latter being a measure of "elasticity" between zero and one and the former being a scale-factor. Suppose a mid-decade census can be taken at cost D, to be borne equally by each local area. If the census is taken, the $\hat{N}_i$ are error free. If the census is not taken, $\Sigma\hat{N}_i$ is error free but each $\hat{N}_i$ has variance $\sigma^2 N^2$ and expectation N. The variance of $\hat{N}_i$ arises because although $\Sigma\hat{N}_i$ is known, it is <u>not known</u> a priori that each area has the same actual population size, and estimates $\hat{N}_i$ must be obtained from imperfect data. Should the census be taken?

To answer this questions in a special case we make the following assumptions (compatible with Redfern):

(i) $\hat{N}_i$ is symmetrically distributed about N for $i = 1,\ldots,n$

(ii) With probability 1, $\hat{N}_i$ lies between $(1-\lambda)N$ and $(1+\lambda)N$ for $i=1,\ldots,n$ and some constant λ between 0 and .1

(iii) $D/M < .1$.

Notice that since $\text{Var}(\hat{N}_i) = N^2\sigma^2$, assumption (ii) implies $\sigma^2 \leq .01$. Under these circumstances the expected increase in social welfare to each local area is

(1.4) $$K(M/n)^{1-B}(1 - B(1-B)\,\sigma^2/2 + (1-B)\,\varepsilon_1)$$ if no census is taken

and

(1.5) $$K(M/n)^{1-B}(1 -(1-B)\,(D/M) - (1-B)\varepsilon_2)$$ if census is taken

where $|\varepsilon_1| < 4 \times 10^{-5}$, $0 < \varepsilon_2 < 6.2 \times 10^{-3}$.

A mid-decade census is justified if

(1.6) $$\frac{D}{M} < \frac{B}{2}\,\sigma^2 - \varepsilon_3$$

where $\varepsilon_3 = \varepsilon_1 + \varepsilon_2$.

Redfern's model is important. Although analysis (below) does not encourage belief that a simple application of his technique can lead to useful results, some bonuses do result from the analysis, such as explicit recognition of the difficulty of the problem of measuring benefits. Furthermore, the distributional assumptions and mathematical analysis are typical of those used in the sequel.

Proof of (1.4) - (1.6)

In the absence of a mid-decade census the expected increase in social welfare to each local area is

$$EK(M\hat{N}_i/\Sigma\hat{N}_j)^{1-B} = K(M/n)^{1-B} \quad E(\hat{N}_i/N)^{1-B} .$$

Expanding $(\hat{N}_i/N)^{1-B}$ in Taylor series about 1 gives (Apostol 1957, p.96)

$$(\hat{N}_i/N)^{1-B} = 1 + (1-B)\cdot(\frac{\hat{N}_i}{N} - 1) - \frac{B(1-B)}{2} \cdot (\frac{\hat{N}_i}{N} - 1)^2$$

$$+ \frac{B(1-B)(1+B)}{6} \cdot (\frac{\hat{N}_i}{N} - 1)^3 + R$$

where $R = \frac{-B(1-B)(1+B)(2+B)}{24} (\frac{\hat{N}_i}{N} - 1)^4 (1+t)^{-3-B}$ for some t

satisfying $|t| < \lambda$. Recall λ was defined by (ii) above.

Since $B(1+B)(2+B) < 6$, we see that

$$|ER| < \frac{(1-B)}{4} \lambda^4 (1-\lambda)^{-4} < \frac{(1-B)}{4}(.1)^4 (.9)^{-4} < (1-B)(4\times 10)^{-5}) .$$

The expected gain in the absence of a census is thus

$$K (M/n)^{1-B} (1-B(\frac{1-B)}{2}\frac{\sigma^2}{2} + (1-B) \varepsilon_1)$$

where $|\varepsilon_1| < 4\times 10^{-5}$.

This establishes (1.4).

If a mid-decade census is taken at cost D, the (expected) increase in social welfare to each local area is $K[(M-D)/n]^{1-B}$. This result follows from (1.3), since the total funds available for distribution is M-D, to be shared equally by each of the n areas.

Expanding in Taylor series about $D=0$ gives

$$(M-D)^{1-B} = M^{1-B} - (1-B)DM^{-B} - \frac{B}{2}(1-B)D^2 (M-\gamma)^{-B-1}$$

where $0 < \gamma < D$. Rewriting this as

$$(M-D)^{1-B} = M^{1-B}[1-(1-B)D/M - (1-B)\varepsilon_2]$$

with $0 < \varepsilon_2 = \frac{B}{2}\frac{D^2}{M^2}(1-\frac{\gamma}{M})^{-B-1} < \frac{1}{2}(.01)(.9)^{-2} < .00618$

and multiplying by Kn^{B-1} we see that the expected gain in social welfare when the census is taken is

$$K(M/n)^{1-B}[1-(1-B)(D/M) - (1-B)\varepsilon_2] .$$

This establishes (1.5). Finally, note that (1.4) < (1.5) if and only if (1.6) holds with $\varepsilon_3 = \varepsilon_1 + \varepsilon_2$.

*****

Redfern's model is not weakened by the likely possibility that other benefits would arise from the mid-decade census data. Other benefits are considered by requiring other uses of the data to bear some of the cost, effectively reducing the magnitude of the parameter D. From the point of view of improved allocations a census for which D/M is less than approximately $B\sigma^2/2$ is a bargain. Actually, Redfern's model is unlikely to provide justification for a mid-decade census. For example if $\sigma = 0.1$ and $D/M = 0.01$ then in (1.6) $B\sigma^2/2 \leq 0.005$ and investment in a mid-decade census is not justified. Note that comparison of costs and benefits of data is performed despite the absence of properties P1 and P2.

It is not obvious how to extend the model to more complicated allocation programs such as GRS. Although Redfern nowhere indicates that he intended such extensions to be possible, attempting to apply his model to GRS illustrates the difficulties of the social welfare-utility theory approach in general. One difficulty arises from considering different size recipient units. Note that true population size does not appear explicitly in (1.3). Because interpersonal utilities are not additive it is not legitimate to write (1.3) as

$$u(y) = NK(y/N)^{1-B}$$

In GRS the true populations (and ideal allocations) are not all equal so it is not clear how to modify $u(.)$ to obtain consistency, in the sense of P3. Finally, it is not known how to approximate the value of B.

Another problem comes from different levels of wealth of recipients in GRS. The usual measures of utility assume that utility from additional income depends upon the wealth of the recipient (e.g. Friedman and Savage 1948). If this assumption is rejected, the utility $u(y)$ for additional income y must have the form (Pfanzagl 1959)

$$u(y) = Ay + C \tag{1.7}$$

or

$$u(y) = AS^y + C \tag{1.8}$$

for some constants A, C, and S. Extending Redfern's model to GRS thus requires incorporating a wealth parameter into (1.3), or replacing (1.3) by (1.7) or (1.8). Even if an appropriate form for $u(.)$ could be found, the complicated structure of GRS makes it likely that spending money to carry out a mid-decade census would increase expected welfare for some recipients and decrease it for others. Thus the difficult problem of comparing welfare of different individuals arises.

§ 1.2 Equity

Equity is an important concern but a vague concept. Congress's concern for equity can be seen from the following passage in the General Revenue Sharing legislation (P.L. 92-512, sec. 109(a)(7)(B))

> ...where the Secretary determines that the data... are not current enough or not comprehensive enough to provide for equitable allocations, he may use... additional data...

Thus the law itself expresses a concern for equity. The courts recently dismissed a challenge by the cities of Newark and Baltimore that the Treasury Department's decision not to adjust their GRS allocations to reflect population undercount in the census was inequitable. The decision underscores how nebulous the term equity is (Newark et al. v. Blumenthal et al., U. S. District Court for the District of Columbia, 1978):

> The term 'equitable allocation' is imprecise, to say the least...There is no standard [of equity] that [the court] could apply except to substitute its judgement for that of the Secretary...[In requiring that data be used to determine the allocations] surely Congress was not attempting to mandate mathematical certainty...At the most, 'equitable allocation' connotes a division that is reasonable, impartial, and fair.

Apparently the courts have no better idea what equitable means than the Secretary of the Treasury.

For our purposes the following non-technical working definition of equity will serve (Black 1968, p. 364):

> Equity. In its broadest and most general signification, this term denotes the spirit and the habit of fairness, justness, and right dealing which would regulate the intercourse of men with men...In a restricted sense, the word denotes equal and impartial justice as between two persons whose rights or claims are in conflict: justice, that is, as ascertained by natural reason or ethical insight, but independent of the formulated body of law.

Our concern with equity is to derive a measure of inequity that we can use as a loss function for deciding how much data quality is appropriate.

There is an intimate connection between equity and utility theory (as described in §1.1 above). Firth (1952) argues that equity can be interpreted as reflecting the preferences of an Ideal Observer, an utterly rational and impartial spectator. Situation A is more equitable than Situation B if and only if the Ideal Observer would prefer A to B from a general point of view. Of course eliciting the preferences* of an idealized being is even more formidable a task than eliciting the preferences of a person. In practice, the method for deriving a loss function from equity considerations usually resembles the utility approach exemplified by Redfern's analysis. The SRI (1974b) study discussed below provides a specific illustration of this method.

The general method for deriving a loss function from equity considerations will now be outlined. A measure of the "inequity to recipient i" from misallocation $f_i - \theta_i$ is defined and denoted by $\Delta_i(\underset{\sim}{f}, \underset{\sim}{\theta})$. A measure $\overline{\Delta}$ of "total inequity" is then derived from the individual Δ_i's. The problem of choosing the appropriate functional form for Δ_i is largely unstudied. With the assumption that the Δ_i are given, derivation of $\overline{\Delta}$ has received considerable attention.

We now consider two derivations of alternative measures $\overline{\Delta}$. These alternative derivations, which we associate with the work of John Rawls and John Harsanyi, are particularly important because the equity-based loss functions in the literature, and also those that we will later construct, use them as a basis.

John Rawls (1971, p.83; see also Rawls 1958) argues that

> Social and economic inequities are to be arranged so that they are...to the greatest benefit of the least advantaged.

---

* under certainty or uncertainty

From the perspective of GRS, if $\underset{\sim}{\theta}$ were to be allocated then each recipient would be equally well-off. Otherwise $\underset{\sim}{\theta}$ would not have been chosen to be the vector of ideal allocations. To see this, suppose each recipient were not equally well-off after $\underset{\sim}{\theta}$ was allocated. Then redistributing some funds from the most well-off to the least well-off would be a more equitable distribution of resources. Such a circumstance violates our basic assumption in §1.0 that $\underset{\sim}{\theta}$ was the optimal vector of allocations. To avoid contradiction we must therefore presume that if $\underset{\sim}{\theta}$ were to be allocated then each recipient would be equally well-off. After allocation $\underset{\sim}{f}$ is made, the least advantaged recipient is the one with the largest inequity, $\Delta_i(\underset{\sim}{f},\underset{\sim}{\theta})$. To arrange the inequities to be of the greatest benefit is thus to minimize the inequity to the recipient with the largest inequity. That is, $\max \{\Delta_i(\underset{\sim}{f},\underset{\sim}{\theta})\}$ should be minimized. This is called the "maximin" rule.

Let $\underset{\sim}{\Delta}'$, $\underset{\sim}{\Delta}''$ denote two vectors of individual inequities with ordered values $\Delta'_{(j)}$, $\Delta''_{(j)}$, i.e.

$$\Delta'_{(1)} \geq \Delta'_{(2)} \geq \ldots \geq \Delta'_{(n)}$$

$$\Delta''_{(1)} \geq \Delta''_{(2)} \geq \ldots \geq \Delta''_{(n)} \quad .$$

The maximin rule says that $\underset{\sim}{\Delta}'$ is preferred to $\underset{\sim}{\Delta}''$ if $\Delta'_{(1)} < \Delta''_{(1)}$. Presumably (Sen 1970, ch.9) if

$\Delta'_{(1)} = \Delta''_{(1)}$ then $\underset{\sim}{\Delta}'$ is preferred to $\underset{\sim}{\Delta}''$ if $\Delta'_{(2)} < \Delta''_{(2)}$.

Generally, $\underset{\sim}{\Delta}'$ is preferred to $\underset{\sim}{\Delta}''$ if, at the first place j where they differ, $\Delta'_{(j)} < \Delta''_{(j)}$. This ordering is called a <u>lexicographic ordering</u>. Unfortunately, for $n > 1$ a lexicographic ordering cannot be represented by a utility function as in (1.1); see Debreu (1965, p.72). Thus the lexicographic "maximin" rule does not provide for a scalar-valued measure of total inequity, and while this rule may be kept in mind, it will be cumbersome to use as an explicit measure of inequity.

Harsanyi (1955, p.316) advocates a weighted sum instead of a maximin rule by interpreting equity as "impersonality":

> An individual's preferences satisfy this requirement of impersonality if they indicated what social situation he would choose if he did not know what his personal position would be in the new situation chosen (and in any of its alternatives) but rather had an equal chance of obtaining any of the social positions existing in this situation, from the highest to the lowest.

Suppose (1) $-\Delta_i(\underset{\sim}{f},\underset{\sim}{\theta})$ is a von Neumann-Morgenstern utility for the preferences of the ith recipient, (2) the preferences are "impersonal", and (3) the Δ_i are scaled so they are comparable (this rather strong assumption is discussed by Harsanyi in section V of his paper). Then it follows that: each individual has the same (impersonal) preference ordering over possible allocations $\underset{\sim}{f}$; this preference ordering is reflected by $\overline{\Delta}(\underset{\sim}{f},\underset{\sim}{\theta}) = \Sigma[\lambda_i \Delta_i(\underset{\sim}{f},\underset{\sim}{\theta})]$, $\lambda_i > 0$; $-\overline{\Delta}$ is a von Neumann-Morgenstern utility. This ordering is the same as that prescribed by utilitarianism (Rawls 1971, p.162).

The orderings above ascribed to Rawls and Harsanyi are two of the most prevalent measures of total inequity appearing in the literature. They can both be derived axiomatically from various "reasonable and desirable" postulates (Sen 1977, especially theorem 5). In both approaches, the functions Δ_i are assumed given.

As the name suggests, Rawls's maximin rule relates closely to the familiar minimax criterion (Lehmann 1959; Luce and Raiffa 1957; Savage 1954). Recall that a minimax rule is one which minimizes the maximum expected loss that can occur under any possible state of nature. Taking an "impersonal" perspective, we may regard the social situation of an individual as having a probability distribution over the actual extant social positions and notice that the minimax and maximin rules coincide, for both rules prescribe minimizing $\max\{\Delta_i(\underset{\sim}{f},\underset{\sim}{\theta})\}$.

The maximin criteria leads to very cautious strategies and while it protects against severe inequity for any individual it may perform poorly for individuals on the average. Repeated over time, the maximin rule can perform badly for all individuals. On the other hand the rule prescribed by Harsanyi may lead to low inequities for individuals on the average but cause severe inequities for certain identifiable population subgroups. It is relevant to note that because blacks are undercounted more than the population as a whole (Siegel et al. 1977) there is concern that areas with high proportions of blacks are consistently being shortchanged in fund allocations (Congress 1977).

A compromise between the maximin criterion of Rawls and the average inequity criterion of Harsanyi retains the advantages of both. The compromise criterion ranks vectors $\underset{\sim}{\Delta}$ of individual inequities lexicographically if any component Δ_i exceeds a level α_i. But if each Δ_i is smaller than α_i, the vectors are ranked according to the average inequity. In application, this compromise rule ensures both that no individual inequity is intolerably high and also that the average inequity is small.

Several attempts to use equity to derive a loss function are of interest. First we consider the work of the Stanford Research Institute (1974b,section V) to evaluate the data base of the GRS program from the perspective of the resulting equity of allocations. The SRI based its measures of inequity on the quantity $f_i/\theta_i - 1$, arguing (ibid, pp.V-9, V-10)

> The definition of equitable allocations as those that would result if all units had accurate, timely data, suggests that the natural measure of equity for a recipient is the ratio of the observed allocation to the ideal allocation. Insofar as this ratio deviates from unity, the recipient concerned has been treated inequitably.
>
> ...the problem of selecting transformations that take into account that equal differences in percents do not necessarily represent equal differences in...equity ...has many conventional solutions, none of which is particularly satisfying. In this analysis, three of these transformations have been adopted.

The three measures of inequity for a recipient were essentially (SRI 1974b, p.V-11)

(1.9) $$\Delta'_i = w_i|f_i/\theta_i - 1|$$

(1.10) $$\Delta''_i = w_i(f_i/\theta_i - 1)^2$$

and

(1.11) $$\Delta'''_i = (w_i/\lambda)\exp[\lambda(1 - f_i/\theta_i)] \qquad \lambda>0$$

where $w_i > 0$ and $\Sigma w_i = 1$. Two sets of weights were considered, $w_i = n^{-1}$ and $w_i = \theta_i/\Sigma\theta_j$.

As functions of $f_i/\theta_i - 1$ the measures (1.9) and (1.10) are non-negative, convex, and symmetric about zero. They are commonly used as loss function in statistical decision theory (see, e.g. Raiffa and Schlaifer 1972). The measure (1.11) is a variant of the "exponential utility function" often used in decision analysis (Raiffa 1968). As a function of f_i/θ_i, (1.11) is asymmetric, strictly decreasing, convex, and of the form (1.8). The parameter λ regulates the amount of convexity. SRI (ibid, p.VI-4) notes that "one of the problems of using... [(1.11)] is the justification for choosing any particular parameter value [for λ]".

The measures of total inequity used by the SRI are all strictly increasing transformations of the sums of (1.9) - (1.11) respectively. Up to translations, they are (Stanford Research Institute 1974b, pp.E-14, V-27):

(1.9') $$\bar{\Delta}' = \Sigma\Delta'_i = \Sigma w_i|f_i/\theta_i - 1|$$

(1.10') $$\bar{\Delta}'' = [\Sigma\Delta''_i]^{1/2} = [\Sigma w_i(f_i/\theta_i - 1)^2]^{1/2}$$

(1.11') $$\bar{\Delta}''' = \lambda^{-1}\log(\Sigma\lambda\Delta'''_i) = \lambda^{-1}\log[\Sigma w_i\exp(\lambda(1-f_i/\theta_i))]$$

These measures would seem to connect with Harsanyi's criterion of impersonality, since for a given measure of inequity, be it (1.9), (1.10), or (1.11), the total inequity of a vector of allocations $\underset{\sim}{f}$ is ordered by the weighted sum of the inequities to the recipients. The strictly increasing transformations were not chosen in consideration of "equity under uncertainty" * but rather to facilitate comparisons between $\overline{\Delta}'$, $\overline{\Delta}''$, and $\overline{\Delta}'''$.

Thomas Jabine (1977) presents an interesting comparison of alternative criteria for measuring total inequity arising in allocation of federal funds. The following model is considered. A fixed pie is allocated to states pro rata to the estimated number of clients having a specified characteristic (such as per capita income below a certain threshold). A state's allocation is shared equally, without error, by all of its clients. Let θ_i and f_i denote the ideal and actual allocations to each client in state i. The problem addressed is how to deploy among the states a fixed amount of sampling resources so as to minimize inequity. Jabine considers two equity criteria

(i) Require $E(f_i - \theta_i)^2$ to be the same for all states (and minimize it)

(ii) Minimize $\sum_i N_i E(f_i - \theta_i)^2$

where N_i is the number of clients in state i.

Criteria (i) and (ii) are respectively Rawl's maximin and Harsanyi's impersonality criterion tailored to Jabine's model. The implications for sampling differ drastically. If the ratio

---

* That is, transformations were not chosen to give to the measures the von Neumann-Morgenstern property (1.2).

of clients to population size is the same for each state, criterion (i) implies approximately equal sample sizes for each state, while (ii) calls for allocating sampling effort in proportion to the square root of the total resident population of the state.

§ 1.3 A Simple Loss Function for Errors in Allocation

To measure the losses from inaccurate allocations we will consider a variety of operational loss functions. These are regarded as approximations to the actual unknown loss function that will be useful to the Decision Maker. To derive these operational loss functions we will first arrive at a measure of "loss to each recipient" from misallocation and then aggregate these individual losses.

The motivating allocation process is General Revenue Sharing (GRS). For GRS we assume that the Decision Maker's preferences reflect Congressional intent. Conceivably Congress is concerned with losses in social welfare arising from inaccurate allocation. However, empirical measurement of these losses is extremely difficult because: (i) the number of recipients is large; (ii) ultimate uses of the funds are hard to trace; (iii) losses are not first order, ie. one recipient's (fiscal) loss is another's gain. Losses from inequity are hard to quantify because Congress's perception of inequity is not understood, largely because Congress is not an individual.

We shall first consider loss functions arising from the assumption that the ratio of the loss to a recipient from misallocation x to the loss from misallocation $-x$ does not vary over recipients. This assumption will be relaxed in §1.7. Recall that in GRS the recipients are state and local governments.

For now assume

$$\text{(1.12)} \qquad \frac{\text{Loss to recipient } i \text{ from misallocation } x}{\text{Loss to recipient } i \text{ from misallocation } -x} = \frac{-b\,H(x)}{a\,H(-x)} \qquad x >$$

where* $a,b > 0$ and H is a real-valued function satisfying

(i) $H(0) = 0$,

(ii) H is strictly decreasing on $(-\infty, 0]$,

and

(iii) H is strictly monotonic on $[0, \infty)$.

Typically, overallocations to a recipient will be regarded as positive benefits to the recipient, so H will be increasing on $[0, \infty)$. It follows from (1.12) that we may express the loss to recipient i from misallocation x_i by

$$w_i\,[\,aH(-x_i^-) \;-bH(x_i^+)\,]\ , \qquad w_i > 0\ ,$$

where $x_i = f_i - \theta_i$ and the functions $(.)^+$, $(.)^-$ are defined by $y^+ = \max(0,y)$, $y^- = \max(0,-y)$. The constants w_i should be thought of as weights allowing that the loss from an underallocation x may be more severe for recipient j than for recipient k , so that $w_j > w_k$. The total loss to recipients is defined to be the sum

$$\sum_i w_i\,[\,aH(-x_i^-) \;-bH(x_i^+)\,]\ .$$

The optimal level of total allocations is $M_\theta = \Sigma\,\theta_i$. Because of errors in data the actual pie distributed, $M_f = \Sigma\,f_i$, may differ from M_θ . If too much money is allocated, that is $M_\theta < M_f$, then the source of the allocations, perhaps the Department of Treasury, the Department of Health, Education and Welfare, or another department or agency, suffers a loss.

---

* It is convenient not to use a single symbol for b/a because we will be interested in $b-a$ as well as b/a and because we will later consider interpretations for a and b .

Similarly, if too little money is distributed, that is $M_\theta > M_f$, then the source may benefit. Treating the source of the allocations analogously to the recipients, we define the loss to the source from misallocation x. by

$$w[a'H(x.^-) \ -b'H(-x.^+)] \qquad w,a',b' > 0$$

where $x. = \Sigma x_i = \Sigma (f_i - \theta_i) = M_f - M_\theta$ and $w,a',b' > 0$.
The total loss from misallocation, say $L_{H,w}(\underset{\sim}{f}, \underset{\sim}{\theta})$, is defined by

$$L_{H,w}(\underset{\sim}{f}, \underset{\sim}{\theta}) = \Sigma w_i \ [aH(-x_i^-) - bH(x_i^+)] + w \ [a'H(x.^-) \ -b'H(x.^+)] \tag{1.13}$$

where $x_i = f_i - \theta_i$; $x. = M_f - M_\theta$; $a,b,a',b',w_i,w > 0$.
The parameters w,w_i can depend on $\underset{\sim}{\theta}$ if $\underset{\sim}{\theta}$ is conceived of as non-random.

Such loss functions are approximations because they are derived from postulates rather than elicitation of the preferences of the policymakers. Usefulness of these loss functions depends on (i) their flexibility to approximate the actual unknown loss function, (ii) the estimability of parameters, so that a good approximation can be obtained, and (iii) their tractability, so that the Decision Maker can use them to compare expected benefits of alternative programs.

With $x_i = f_i - \theta_i$, $x. = M_f - M_\theta$, and $a,b,a',b',w,w_i > 0$ we have the following special cases of (1.13):

$$L_H(\underset{\sim}{f}, \underset{\sim}{\theta}) = \sum_i w_i[aH(x_i^-) \ -bH(x_i^+)] + w[a'H(x.^+) \ -b'H(x.^-)] \tag{1.14}$$

where $H(0) = 0$ and H is strictly increasing on $[0, \infty)$;

$$L_w(\underset{\sim}{f}, \underset{\sim}{\theta}) = \sum_i w_i[ax_i^- \ -bx_i^+] + w[a'x.^+ \ -b'x.^-] \tag{1.15}$$

and

$$(1.16) \qquad L(\underset{\sim}{f}, \underset{\sim}{\theta}) = \sum_i [a x_i^- - b x_i^+] + a' x.^+ - b' x.^- \quad .$$

The loss function L_H is obtained by requiring $H(x) = H(-x)$ in (1.13). The loss function L_w is a special case of L_H with $H(x) = x$, $x > 0$. Setting all weights w, w_i to unity in L_w yields L .

The remainder of this section considers only $L(\underset{\sim}{f}, \underset{\sim}{\theta})$. The loss functions $L_H(\underset{\sim}{f}, \underset{\sim}{\theta})$, $L_w(\underset{\sim}{f}, \underset{\sim}{\theta})$ are analyzed in §1.6. Loss functions with non-symmetric H will be considered in §1.7 in a setting less restrictive than (1.12).

The model for loss to a recipient under $L(\underset{\sim}{f}, \underset{\sim}{\theta})$ may be represented

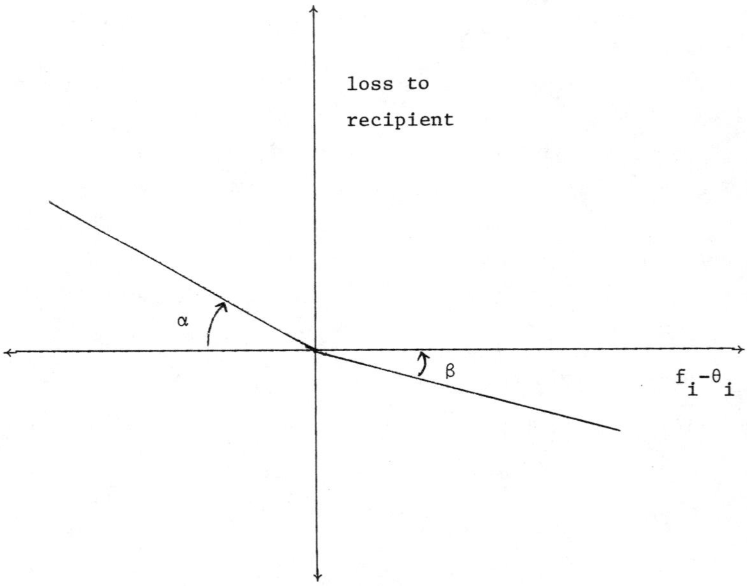

where $\dfrac{\tan \beta}{\tan \alpha} = \dfrac{b}{a}$.

Because of the convexity of most utility functions, one feels that b/a is less than 1, but not too small. More understanding of the constants a and b can be gotten from the following idealized scenario. Suppose the θ_i are such that an allocation less than θ_i causes a deficit to recipient i while an allocation greater than θ_i produces a surplus. If $f_i < \theta_i$ so that a deficit is incurred, recipient i is assumed to borrow $\theta_i - f_i$. The loan is repaid in the next fiscal period, say one year later. If the yearly interest rate was a-1, the loss in dollars to recipient i would be $a(\theta_i - f_i)$. We neglect long term effects because they are hard to trace and because the recipient cannot make adjustments for the deficit in less than a year's time, but after a year adjustments can be made. Conversely, if $f_i > \theta_i$ so that a surplus is produced, recipient i invests $f_i - \theta_i$ for a year at interest rate b-1. The loss incurred in this case is $-b(f_i - \theta_i)$. Generally, the loss to recipient i can be written $a(f_i - \theta_i)^- - b(f_i - \theta_i)^+$. Because the losses are in dollars it is legitimate and meaningful to sum them to obtain the total loss to recipients from misallocation:

$$\Sigma\, [a(f_i - \theta_i)^- - b(f_i - \theta_i)^+] \quad . \tag{1.17}$$

The constants a' and b' may be interpreted in the same manner to yield the loss to the source:

$$a'(M_\theta - M_f)^- - b'(M_\theta - M_f)^+ \quad . \tag{1.18}$$

By summing (1.17) and (1.18) we obtain (1.16).

Use of interest rates provides an upper bound to the loss. For example, in GRS recipient i will be a state or local government and the option of cutting back services might be chosen in preference to borrowing money. From the assumption of rational behavior on the part of the government this revealed preference indicates that in this case a larger loss would be associated with borrowing.

Similarly, $-b(f_i - \theta_i)^+$ is an upper bound to the loss incurred by the recipient from overallocation. We are assuming here that the Congress's evaluation of the loss takes into account that of the recipient.

Notice that the constants a,b,a',b' can be adjusted to reflect losses from inequity as well. If we define the <u>inequity to recipient i</u> by $f_i - \theta_i$, then we can decompose this inequity into the sum of the <u>horizontal inequity</u> (or <u>distributive inequity</u>) $M_f(f_i/M_f - \theta_i/M_\theta)$ and the <u>vertical inequity</u> $(M_f - M_\theta) \cdot (\theta_i/M_\theta)$. The vertical inequity to a recipient is the burden of the error in the size of the allocated pie that the recipient i would bear if there were no error in the determination of the recipient's <u>share</u> of the pie. If the actual pie is the correct size then there is no vertical inequity. The net vertical inequity (vertical inequity summed over all recipients) always equals the error in the total pie distributed. The horizontal inequity to a recipient is the actual total burden of error in the actual share of the pie for the recipient. If the actual share of the pie is correct, the horizontal inequity is zero, regardless of whether there is any vertical inequity. The net horizontal inequity is always zero. Treating the actual allocations as random variables, we can easily show that the vertical and horizontal inequities to recipient i are uncorrelated when $E(f_i/M_f | M_f) = \theta_i/M_\theta$.

Note that there is some apparent arbitrariness in the above definitions of vertical and horizontal inequities. The weight for the vertical component, θ_i/M_θ, is based on the optimal allocations whereas the weight for the horizontal component, M_f, is based on the actual allocations. The definitions could alternatively have been vertical inequity equals $(M_f - M_\theta)(f_i/M_f)$ and horizontal inequity equals $M_\theta(f_i/M_f - \theta_i/M_\theta)$. The two components still sum to the total inequity for recipient i and allow interpretations similar to those of the components as originally defined. However, these alternatively defined components are <u>not</u> uncorrelated, and for this reason I prefer the original definitions.

§ 1.4 Estimating the Parameters of the Loss Function $L(\underset{\sim}{f}, \underset{\sim}{\theta})$ for GRS

The values of a,b,a',b' in (1.16) reflect the Decision Maker's perception of the sensitivity of the policymakers to errors in allocation. Restricting attention to GRS, we will demonstrate that although precise estimates of these constants are elusive, useful bounds are possible.

First note that when $M_f = M_\theta$, $L(\underset{\sim}{f}, \underset{\sim}{\theta})$ equals

(1.20) $$\frac{a-b}{2} \; \Sigma \, |f_i - \theta_i| \quad .$$

Since GRS was essentially a fixed-pie allocation process from 1972 to 1976 we will concentrate on bounds for a,b. As it is unlikely that Congress is so sensitive to errors in GRS allocations that the loss could exceed $\Sigma \, |f_i - \theta_i|$ or even $\Sigma \, (f_i - \theta_i)^-$, values of $(a-b)/2$ as big as 1 or even only 1/2 might be deemed "unreasonably large".

The interest rate interpretation also gives some feeling for appropriate values of a,b. For example, it seems reasonable that $(a-b)/2$ is no greater than 0.05 and possibly $(a-b)/2$ is in the neighborhood of 0.001. It is also plausible that the ratio of b to a is less than 1 but not by much, say

(1.21) $$.75 < b/a < 1 \quad .$$

<u>Example 1.1</u>

Let $f_i(t)$ and $\theta_i(t)$ denote the actual and optimal GRS allocations to recipient government i in Entitlement Periods (EP's) $t = 1,2 \ldots T$. Let B denote the amount of money spent on determining allocations after the first EP and denote by $D(B)$ the expected reduction in absolute misallocations from expenditure B. Since GRS is essentially fixed pie we will write

$$L(\underset{\sim}{f}(t)\ ,\ \underset{\sim}{\theta}(t)) = \frac{a-b}{2} \sum_i | f_i(t) - \theta_i(t) | \qquad \text{for EP } t$$

and

$$L(\underset{\sim}{f}, \underset{\sim}{\theta}) = \frac{a-b}{2} \sum_t \sum_i |f_i(t) - \theta_i(t)| \ .$$

If no money had been spent on determining allocations after the first EP, then the estimates $f_i(1)$ would continue to be used and the expected sum of absolute misallocations would be $E \sum_t \sum_i |f_i(1) - \theta_i(t)|$. Thus we see that

$$D(B) = E(\Sigma\Sigma|f_i(1) - \theta_i(t)| - \Sigma\Sigma|f_i(t) - \theta_i(t)|)$$

$$\leq E(\Sigma\Sigma|f_i(1) - f_i(t)|) \ .$$

Suppose $B^* \neq 0$ was the optimal amount to spend on data and analysis to obtain the estimates $f_i(t)$. Then† $B^* = (a-b)\cdot D(B^*)/2$ and so

$$(1.23) \qquad (a-b)/2 \geq B^* / E(\Sigma\Sigma|f_i(1) - f_i(t)|) \ .$$

---

† To show that $B^* = (a-b)\ D(B^*)/2$, let $K = (a-b)/2$, let $Y(B)$ = sum of absolute misallocations when B is spent, and assume D is differentiable. Then

$$\begin{aligned} \text{Expected loss from spending } B &= B + K\ EY(B) \\ &= B + K\ E(Y(B) - Y(0) + Y(0)) \\ &= B - KD(B) + KY(0) \ . \end{aligned}$$

Since $B^* \neq 0$ it follows that

$$0 = 1 - KD'(B^*) \ ,$$

which implies $B^* = KD(B)$.

If EP T has already occurred then $f_i(t)$ is not random and so $\Sigma\Sigma\ |f_i(t) - f_i(1)|$ is easily evaluated. Although the optimal value for B will not be known, the actual value of B can be found. Substituting this actual value for B^* in (1.23) provides a lower bound for $(a-b)/2$ that is useful for two purposes:

(i) Retropsective analysis of past expenditures for data collection (including processing) in GRS. For example, the lower bound might suggest that $(a-b)/2$ was unreasonably large, so that too much had been spent on data collection.

(ii) Comparing current or future expenditures for data with past levels of expenditure. For example, the decision of whether or not to take a mid-decade census could be examined for consistency with previous data expenditure decisions.

Empirical investigation of the bounds on $(a-b)/2$ will not be undertaken in this research. In the sequel (chapters 6 and 7 especially) the absolute error loss function will be used with $(a-b)/2 = .01$.

The remainder of this chapter treats theoretical aspects of loss functions. The material is important for choosing a loss function and is interesting in its own right but is not used in the sequel. The reader primarily interested in the benefit-cost applications is encouraged to skip to chapter 3, at least for a first reading.

§ 1.5 Fisher Consistency and Other Properties of the Loss Function $L(\underset{\sim}{f}, \underset{\sim}{\theta})$

Various mathematical and statistical properties of $L(\underset{\sim}{f}, \underset{\sim}{\theta})$ are presented below. The important criterion of Fisher-consistency will also be developed and applied to loss functions in the literature. First note

Lemma 1.1 $L(\underset{\sim}{f}, \underset{\sim}{\theta})$ as defined by (1.16) may be written

$$(1.26) \quad L(\underset{\sim}{f}, \underset{\sim}{\theta}) = \frac{a-b}{2} \Sigma|f_i - \theta_i| + \frac{a' - b'}{2} | M_f - M_\theta| + (\frac{a' + b'}{2} - \frac{a+b}{2})(M_f - M_\theta) .$$

<u>Proof</u>:

Let $x_i = f_i - \theta_i$ and $x. = \Sigma x_i$. From (1.17) the sum of the losses to recipients is

$$a(\Sigma x_i^-) - b(\Sigma x_i^+)$$

$$= (a/2)(\Sigma|x_i| - \Sigma x_i) - (b/2)(\Sigma|x_i| + \Sigma x_i)$$

$$= \frac{(a-b)}{2}(\Sigma|x_i|) - \frac{(a+b)}{2} x. \quad .$$

The loss to the source is

$$a' x.^+ - b' x.^-$$

$$= (a'/2)(|x.| + x.) - (b'/2)(|x.| - x.)$$

$$= \frac{(a' - b')}{2} |x.| + \frac{(a' + b')}{2} x. \quad .$$

The total loss is obtained by summing. Thus

$$L(\underset{\sim}{f}, \underset{\sim}{\theta}) = \frac{(a-b)}{2} \Sigma|x_i| + \frac{(a' - b')}{2} |x.|$$

$$+ \left[\frac{(a' + b')}{2} - \frac{(a+b)}{2}\right] x. \quad ,$$

which translates into (1.26).

*****

Corollary 1.2 If $a' = a$ and $b' = b$ in (1.16) then

$$L(\underset{\sim}{f},\underset{\sim}{\theta}) = \frac{(a-b)}{2}\left[\,\Sigma|f_i - \theta_i| + |M_f - M_\theta|\,\right] \tag{1.27}$$

$$= (a-b)\cdot \max\{\,\Sigma(f_i - \theta_i)^+ \;,\; \Sigma(f_i - \theta_i)^-\,\}\,.$$

Some configurations of the constants a,b,a',b' have pathological implications. For example, if $a = b < a' = b'$ then from (1.26)

$$L(\underset{\sim}{f},\underset{\sim}{\theta}) = K\,(M_f - M_\theta) \qquad \text{where } K = a' - a > 0 \;. \tag{1.28}$$

In this case the optimal estimator M_f does not depend on the data but takes as small a value as possible (conceivably $-\infty$) .

Such a loss function is considered by Jabine and Schwartz (1974, p.103) in the context of determining optimal sample size for data collection in the Supplemental Security Income program (SSI). SSI allocates funds $\hat{A}_i$ from the federal government to the states $i = 1,\ldots 51$ (including the District of Columbia) according to sample estimates $\hat{Y}_i$ of population parameters Y_i . Let A_i denote the allocation when $\hat{Y}_i = Y_i$. Jabine and Schwartz consider (p.104) that

> ...we represent the Federal government, and we want to minimize losses from estimates which result in <u>overpayments</u> to the States. We may then define a payment error function
> .

$$L_3 = \hat{A}_i - A_i \quad \text{when} \quad \hat{A}_i \geq A_i \tag{1.29}$$

$$= 0 \qquad \text{otherwise.}$$

Note that the loss function L_3 has the same pathology as (1.28). The optimal estimator for A_i is as small as possible and does not depend upon the data. To get around this Jabine and Schwartz only consider estimators $\hat{A}_i$ such that $E\,\hat{A}_i = A_i$. This is an irrelevant* constraint, imposed merely to ensure that the optimal $\hat{A}_i$ wil depend on data in some way. The real problem is that (1.29) is not an appropriate loss function for their problem.

In general, a loss function for the problem of estimating $\underset{\sim}{\theta}$ by $\underset{\sim}{f}$ is called <u>Fisher-consistent</u>** if and only if it takes its unique minimum value when $\underset{\sim}{f} = \underset{\sim}{\theta}$. In particular, loss functions (1.28), (1.29) are not Fisher-consistent. Lack of <u>Fisher-consistency</u> can lead to pathologies such as Jabine's and the following example.

<u>Example 1.2</u>

The loss function corresponding to the measure of inequity (1.11') used by the SRI is not in general Fisher-consistent. Writing this loss function as

$$L_S(\underset{\sim}{f}, \underset{\sim}{\theta}) = \lambda^{-1} \log[\, \Sigma\, w_i \exp(\, \lambda\, (1 - f_i/\, \theta_i))] \ ,$$

$$\text{where} \quad \lambda,\, w_i > 0 \quad \text{and} \quad \Sigma\, w_i = 1 \quad .$$

We note that for given positive vectors $\underset{\sim}{\theta}$, $\underset{\sim}{f}$ there always exist choices of $\lambda > 0$ such that $L_S \geq 0$ (and $L_S = 0$ when $\underset{\sim}{f} = \underset{\sim}{\theta}$) .

* Irrelevant from the viewpoint of statistical decision theory. We are treating the problem of determining the allocations as an estimation problem. The approach of statistical decision theory compares alternative statistical procedures solely on the basis of expected loss. Unless the class of procedures under consideration is limited a priori for compelling reasons, constraints such as unbiasedness may be useful mathematical expedients but are not relevant for choosing the "best" procedure, where best is determined on the basis of expected loss.

** Motivation for the terminology <u>consistent</u> comes from Fisher's principle of consistency in estimation theory (Fisher 1959, pp.143-4), which is "essentially a means of stipulating that the process of estimation is directed to the particular parameter under discussion, and not to some other function of the adjustable parameter or parameters".

In fact, Stanford Research Institute (1974b, p.VI-5) chose λ so that the observed values of L_S would be comparable to other measures of loss (specifically (1.9'), (1.10'). SRI (1974b, p.E-12) considered two choices for weights: (1) $w_i = 1/n$ and (2) $w_i = \theta_i/M_\theta$. It will be shown below in §1.7 that L_S is Fisher-consistent if and only if w_i is proportioned to θ_i . Thus the second choice of weights is valid but a loss function with $w_i = 1/n$ is not Fisher-consistent. This explains why SRI (1974b, table V-6, p.V-37) found that anticipated errors in fiscal year 1977-78 and 1978-79 data produced a net <u>gain</u> in equity.

We now consider the Fisher-consistency of $L(\underset{\sim}{f},\underset{\sim}{\theta})$ given by (1.16).

<u>Proposition 1.3</u> If f is a fixed-pie allocation formula such that $M_f = M_\theta$ but $\underset{\sim}{f}$ is not identically equal to $\underset{\sim}{\theta}$ then

(1.30) $L(\underset{\sim}{f},\underset{\sim}{\theta})$ is Fisher-consistent if and only if $a > b$.

<u>Proof</u>: From (1.20) notice $L(\underset{\sim}{f},\underset{\sim}{\theta}) = \frac{(a-b)}{2} \Sigma|f_i - \theta_i|$, which has a unique minimum if and only if $a > b$ and attains its unique minimum, 0 , if and only if $\underset{\sim}{f} = \underset{\sim}{\theta}$.

*****

For a non-fixed pie allocation formula f , proposition 1.4 below characterizes Fisher-consistency for $L(\underset{\sim}{f},\underset{\sim}{\theta})$ under the following regularity condition:

(1.31) There exist data points $\underset{\sim}{z}_1$, $\underset{\sim}{z}_2$, $\underset{\sim}{z}_3$ such that

$M_f(\underset{\sim}{z}_1) = M_\theta$ but for some i $f_i(\underset{\sim}{z}_1) \neq \theta_i$,

$f_i(\underset{\sim}{z}_2) \geq \theta_i$ for all i with strict inequality for some i ,

and

$f_i(\underset{\sim}{z}_3) \leq \theta_i$ for all i with strict inequality for some i .

<u>Proposition 1.4</u> If

(1.32) $a > \max(b,b')$ and $a' > b$

then $L(\underset{\sim}{f},\underset{\sim}{\theta})$ is Fisher-consistent. Furthermore, if $L(\underset{\sim}{f},\underset{\sim}{\theta})$ is Fisher-consistent and (1.31) holds then (1.32) holds.

<u>Proof</u>: First note $L(\underset{\sim}{f},\underset{\sim}{\theta}) = 0$ when $\underset{\sim}{f} = \underset{\sim}{\theta}$. Now suppose (1.32) holds and $\underset{\sim}{f} \neq \underset{\sim}{\theta}$. From (1.26) if $M_\theta = M_f$ and $\underset{\sim}{f} \neq \underset{\sim}{\theta}$ then

$$L(\underset{\sim}{f},\underset{\sim}{\theta}) = \frac{a-b}{2} \Sigma|f_i - \theta_i| > 0$$

while if $M_\theta > M_f$ then

$$L(\underset{\sim}{f},\underset{\sim}{\theta}) \geq \frac{a-b}{2} |M_f - M_\theta| + \frac{a' - b'}{2} |M_f - M_\theta|$$
$$- (\frac{a' + b'}{2} - \frac{a + b}{2}) \cdot |M_f - M_\theta|$$
$$= (a-b') |M_f - M_\theta| > 0$$

and if $M_\theta < M_f$ then

$$L(\underset{\sim}{f},\underset{\sim}{\theta}) \geq \frac{a-b}{2} |M_f - M_\theta| + \frac{a' - b'}{2} | M_f - M_\theta|$$
$$+ (\frac{a' + b'}{2} - \frac{a+b}{2}) | M_f - M_\theta|$$
$$= (a' - b) |M_f - M_\theta| > 0 .$$

Thus $L(\underset{\sim}{f},\underset{\sim}{\theta})$ is Fisher-consistent. Now suppose that $L(\underset{\sim}{f},\underset{\sim}{\theta})$ is Fisher-consistent and (1.31) holds. From (1.26) we obtain the following three relations:

(1.33a) $$L(\underset{\sim}{f}(\underset{\sim}{z}_1),\underset{\sim}{\theta}) = \frac{a-b}{2} \quad \Sigma|f_i(\underset{\sim}{z}_1) - \theta_i|$$

(1.33b) $$L(\underset{\sim}{f}(\underset{\sim}{z}_2),\underset{\sim}{\theta}) = \left(\frac{a-b}{2} + \frac{a' - b'}{2} + \frac{a' + b'}{2} - \frac{a+b}{2}\right)(M_f(\underset{\sim}{z}_2) - M_\theta)$$
$$= (a' - b)\ (M_f(\underset{\sim}{z}_2) - M_\theta)$$

and

(1.33c) $$L(\underset{\sim}{f}(\underset{\sim}{z}_3),\underset{\sim}{\theta}) = \left(\frac{a-b}{2} + \frac{a' - b'}{2} + \frac{a+b}{2} - \frac{a' + b'}{2}\right) |M_f(\underset{\sim}{z}_3) - M_\theta|$$

$$= (a - b')\ |M_f(\underset{\sim}{z}_3) - M_\theta| \ .$$

Fisher consistency of $L(\underset{\sim}{f},\underset{\sim}{\theta})$ implies that each of $L(f(\underset{\sim}{z}_1),\underset{\sim}{\theta})$, $L(f(\underset{\sim}{z}_2),\underset{\sim}{\theta})$, and $L(f(\underset{\sim}{z}_3),\underset{\sim}{\theta})$ is strictly positive, which implies that each of $a-b$, $a'-b$, $a-b'$ is strictly positive.

*****

In terms of interest rates, proposition 1.4 says that the loss function $L(\underset{\sim}{f},\underset{\sim}{\theta})$ is Fisher-consistent if and only if

(i) a recipient cannot borrow money at a lower interest rate than his rate of return on investment $(a > b)$

(ii) the source cannot borrow money at a lower interest rate than the recipient's rate of return on investment $(a' > b)$

and

(iii) a recipient cannot borrow money at a lower interest rate than the source's rate of return on investment $(a > b')$.

Failure of (i), (ii), or (iii) introduces instability into the model. If (i) fails, the recipients borrow and then invest continually to make an unlimited profit. If (ii) (or iii) fails, the source (recipients) will borrow money and lend it to the recipients (source), who will invest it at a higher rate to make unlimited net profit for the system. The conditions (i) - (iii) thus serve to close the model, i.e. prevent unlimited input of money.*

If $L(\underset{\sim}{f},\underset{\sim}{\theta})$ is Fisher-consistent then its minimum value is zero and so $L(\underset{\sim}{f},\underset{\sim}{\theta})$ is a regret function (Savage 1972). If $L(\underset{\sim}{f},\underset{\sim}{\theta})$ is not Fisher-consistent, the following corollary shows it cannot be a regret function.

Corollary 1.5 Assuming that $\underset{\sim}{f}$ may take any values in R^n, if $L(\underset{\sim}{f},\underset{\sim}{\theta})$ is not Fisher-consistent then $L(\underset{\sim}{f},\underset{\sim}{\theta})$ is not bounded below.

Proof: By proposition 1.4, one of the inequalities in (1.32) fails. Let m denote an arbitrary positive number. Suppose $a' < b$. Let $f_i = m/n$ for all i. For sufficiently large m, (1.33b) shows that $L(\underset{\sim}{f},\underset{\sim}{\theta}) = (a' - b)(m - M_\theta)$ which is not bounded below. If $a < b'$ let $f_i = m/n$ for all i. For sufficiently large m, (1.33c) shows that $L(\underset{\sim}{f},\underset{\sim}{\theta}) = (a - b')(m - M_\theta)$ which is not bounded below. Finally, if $a < b$ let $f_1 = m$, $f_2 = -m$, and $f_3 = \ldots = f_n = 0$. From (1.26)

$$L(f,\theta) = \frac{a-b}{2}\left(|m - \theta_1| + |m + \theta_2| + \Sigma|\theta_i| \right) + \frac{a' - b'}{2}|M_\theta| - \left(\frac{a' + b'}{2} - \frac{a+b}{2}\right)M_\theta$$

$$= (a-b)m + \frac{a-b}{2}\left(\theta_2 - \theta_1 + \Sigma|\theta_i| \right) + \left(\frac{a+b}{2} - b'\right)M_\theta^{+} + \left(\frac{a+b}{2} - a'\right)M_\theta^{-}$$

which is not bounded below.

*****

---

* The scenario described in §1.3 unfortunately fails to satisfy (i). A quirk in the tax laws allows state and local governments to sell tax-exempt bonds, effectively realizing a lower interest rate for borrowing than investing. Additional mechanisms would therefore have to be introduced for the scenario to be applicable to GRS.

§ 1.6 More General Loss Functions: $L_H(\underset{\sim}{f},\underset{\sim}{\theta})$, $L_w(\underset{\sim}{f},\underset{\sim}{\theta})$

The loss function $L(\underset{\sim}{f},\underset{\sim}{\theta})$ does not take into account different sizes of recipients. Applied to GRS state allocations, it treats unit misallocations to large and small states the same way. The loss function $L_w(\underset{\sim}{f},\underset{\sim}{\theta})$ generalizes $L(\underset{\sim}{f},\underset{\sim}{\theta})$ by weighting the effects of misallocations to recipients. The weights may vary for different recipients but for each recipient overallocations and underallocations are weighted the same. Recall that

$$(1.15) \qquad L_w(\underset{\sim}{f},\underset{\sim}{\theta}) = w_i[a(f_i - \theta_i)^- - b(f_i - \theta_i)^+] + w[a'(M_\theta - M_f)^- - b'(M_\theta - M_f)^+]$$

and

$$(1.14) \qquad L_H(\underset{\sim}{f},\underset{\sim}{\theta}) = \Sigma\, w_i[aH((f_i - \theta_i)^-) - bH((f_i - \theta_i)^+)] + w[a'H((M_\theta - M_f)^-) - b'H((M_\theta - M_f)^+)] \ .$$

The loss function L_H generalizes L_w by transforming the measures $(f_i - \theta_i)^-$, $(f_i - \theta_i)^+$ with an arbitrary increasing function H defined on $[0 , \infty)$ such that $H(0) = 0$; for example $H(x) = x^2$ or $H(x) = e^x - 1$.

It will be shown that requiring L_H to be Fisher-consistent implies that L_H cannot be very different from L_w . Fisher-consistency also implies that the weights w_i can vary only slightly. For the initial GRS state allocations, Fisher-consistency of L_w or L_H is sacrificed by setting $w_i = \theta_i^{-1}$, $\theta_i^{-1/2}$, P_i^{-1} , or $P_i^{-1/2}$ where P_i is the population of state i .

To study Fisher-consistency of $L_H(\underset{\sim}{f},\underset{\sim}{\theta})$ and $L_w(\underset{\sim}{f},\underset{\sim}{\theta})$ we use a strengthened form of regularity condition (1.31):

(1.35) $(\underset{\sim}{f}-\underset{\sim}{\theta})^T$ can assume the values $(x,0,\ldots 0)$, $(0,x,\ldots 0)$, $\ldots(0,\ldots 0,x)$;

$(x, -x,0,\ldots 0)$, $(x,0,-x,0\ldots 0)$, $\ldots(x,0,\ldots 0,-x)$, $\ldots(0,\ldots 0,x,-x)$; and

$(x,x,\ldots x)$ for all real-valued x .

First we note that the weights $w,w_1\ldots,w_n$ in (1.14) or (1.15) cannot vary much.

<u>Proposition 1.6</u>

If the loss function L_H defined by (1.14) is Fisher-consistent and if regularity condition (1.35) holds, then

$$\text{(1.36)} \qquad \frac{\max\{w_i\}}{\min\{w_i\}} < \frac{a}{b} ,$$

$$\frac{\max\{w_i\}}{w} < \frac{a'}{b} ,$$

and

$$\frac{w}{\min\{w_i\}} < \frac{a}{b'} .$$

<u>Proof</u>: Without loss of generality assume $w_1 \le w_2 \le\ldots\le w_n$. The inequalities are proved by requiring $L_H(\underset{\sim}{f},\underset{\sim}{\theta}) > 0$ when $(\underset{\sim}{f}-\underset{\sim}{\theta})^T = (-x,0,\ldots,0,x)$, $(0,\ldots,0,x)$, and $(x,0,\ldots,0)$ for $x > 0$. The method of proof is just like that of proposition 1.4. *****

Implications of Fisher-consistency for L_w and L_H will now be considered. Inequality (1.36) may be rewritten as

$$1 \ge \min\{w_i\} / \max\{w_i\} > b/a .$$

If the w_i are set inversely proportional to θ_i , $\theta_i^{1/2}$, P_i , or $P_i^{1/2}$ then this inequality will fail to hold in applications.

For example, in the initial GRS allocation to states

$$\min \{\theta_i^{-1}\}/\max \{\theta_i^{-1}\} < .02 \ , \quad \min \{\theta_i^{-1/2}\}/\max \{\theta_i^{-1/2}\} < .15 \ .$$

In §1.4 however, it was reasoned that b/a would be much larger than .15 , say $b/a > 0.75$. Thus the weights w_i in L_w and L_H cannot vary significantly.

It is important that the loss function be flexible to handle varying numbers of recipients. For example, the number of recipients in the General Revenue Sharing (GRS) program changed over time as some previously ineligible places became self-governing and therefore newly eligible for GRS allocations. The number of recipients can also vary across different allocation programs or within different parts of one allocation program (e.g. there are more counties in Texas than in New Jersey to receive GRS allocations).

Consequently, Fisher-consistency of a loss function should not depend on the number n of recipients. Note that the constants a, b, a', b' do not vary with n although the set of weights w_i can. The following theorem shows that if L_H is Fisher-consistent for any number of recipients, then $H(x)/x$ is uniformly bounded in n for $x \geq 1$.

Theorem 1.7 If the loss function $L_H(\underset{\sim}{f}, \underset{\sim}{\theta})$ is Fisher-consistent and regularity condition (1.35) holds for vectors $\underset{\sim}{f}$ and $\underset{\sim}{\theta}$ of arbitrary length n , then there exist positive constants B and A not depending on n such that

$$(1.37) \qquad B < \frac{H(x)}{x} < A \qquad x \geq 1 \ .$$

Proof:

For $\underset{\sim}{f} - \underset{\sim}{\theta} = (x, \ldots, x)^T$ with $x > 0$, Fisher-consistency implies

$$a'wH(nx) - b(\Sigma \, w_i)H(x) > 0 \ .$$

Likewise, for $\underset{\sim}{f}-\underset{\sim}{\theta} = -(x,\ldots,x)^T$ with $x > 0$, Fisher-consistency implies

$$a(\Sigma\, w_i)H(x) - b'wH(nx) > 0 \quad .$$

These two inequalities jointly imply

$$\frac{b\ \Sigma w_i}{a'w} < \frac{H(nx)}{H(x)} < \frac{a\ \Sigma w_i}{b'w} \quad .$$

Now use (1.36) to observe

$$\frac{b\ \Sigma w_i}{a'w} \geq \frac{b\cdot n\cdot \min\ \{w_i\}}{a'w} > n\,\frac{b}{a'}\cdot\frac{b'}{a}$$

and

$$\frac{a\ \Sigma w_i}{b'w} \leq \frac{a\cdot n\cdot \max\ \{w_i\}}{b'w} < n\,\frac{a}{b'}\cdot\frac{a'}{b} \quad .$$

Setting $B' = \frac{b}{a}\,\frac{b'}{a'}$ and $A' = 1/B'$ yields

(1.38) $$B' < \frac{H(nx)}{nH(x)} < A' \quad .$$

It is important to note that (1.38) holds for all positive integers n, and also that B' and A' do not depend on n.

For $1 \leq n \leq r < n+1$, the monotonicity of H implies

(1.39) $$\frac{H(nx)}{(n+1)H(x)} < \frac{H(rx)}{rH(x)} < \frac{H((n+1)x)}{nH(x)} \ , \qquad x > 0 \quad .$$

Notice (by 1.38)

(1.40) $$\frac{B'n}{n+1} < \frac{H(nx)}{(n+1)H(x)}$$

and

$$(1.41) \qquad \frac{H((n+1)x)}{nH(x)} < \frac{(n+1)A'}{n}$$

and apply these two inequalities to (1.39) to get

$$(1.42) \qquad \frac{n}{n+1} B' < \frac{H(rx)}{rH(x)} < \frac{n+1}{n} A' .$$

The conclusion follows immediately from the substitutions

$$x = 1 \quad , \quad B = \frac{B'}{2} H(1) \quad , \quad \text{and} \quad A = 2A'H(1) \quad .$$

*****

Theorem 1.7 implies that for $x > 1$ $H(x)$ must must lie in the shaded region of the following picture:

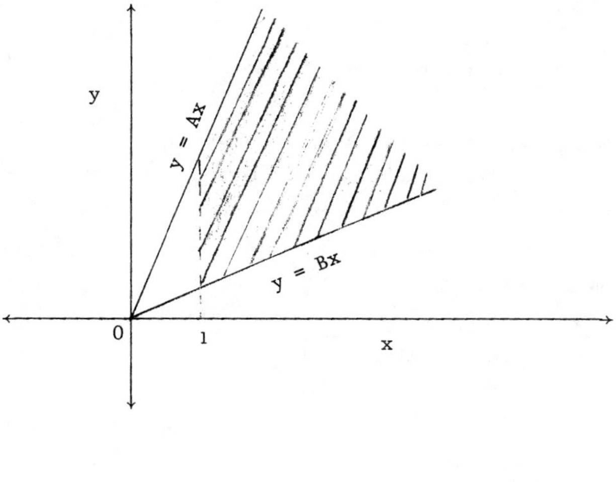

Theorem 1.7 does <u>not</u> imply that H is asymptotically linear; that is H may fail to be of the form $cx + \varepsilon_x$ for some non-zero c and ε_x satisfying $\lim_{x\to\infty} \varepsilon_x / X = 0$ because there exist functions that oscillate eternally between the upper and lower bounds. Such a function appears as example 1.3 in the technical appendix.

Under the regularity condition

(1.43) $\psi(n) = \lim_{x\to\infty} \frac{H(nx)}{H(x)}$ exists for all positive integers n

it will be shown that H satisfies a weaker kind of "asymptotic linearity":

(1.44) $\lim_{x\to\infty} \frac{H(rx)}{H(x)} = r$ for all $r > 0$.

<u>Corollary 1.8</u>

Under the hypotheses of theorem 1.7 and regularity condition (1.43), the function H satisfies (1.44).

<u>Proof</u>: see technical appendix. *****

I conjecture that in fact under these hypotheses
$H(x) = Ax + \varepsilon_x$.

The preceding analysis shows that if a loss function of the form $L_H(\underset{\sim}{f},\underset{\sim}{\theta})$ is to be used, then the loss function $L_W(\underset{\sim}{f},\underset{\sim}{\theta})$ is a good choice because it is simpler to use and not very different from other possible choices. For similar reasons, the loss function $L(\underset{\sim}{f},\underset{\sim}{\theta})$ is a good choice if a loss function of the form $L_W(\underset{\sim}{f},\underset{\sim}{\theta})$ is to be used. The lack of choice in the variation of the weights can however be thought of as a shortcoming of this class of loss functions.

The next section discusses loss functions of the form (1.13) but not of the "symmetric" form (1.14), as well as loss functions of a more general form than (1.13). For these loss functions significant weighting schemes will be possible.

§ 1.7 Exponential Loss Functions

The loss functions considered up to now have mainly been of the "symmetric" form (1.14). We will now consider loss functions of the form (1.13) where H is not symmetric. With some gain in generality we also weaken condition (1.12) and only require the existence of positive constants c_i such that

(1.45) $$\frac{\text{loss to recipient } i \text{ from misallocation } x}{\text{loss to recipient } i \text{ from misallocation } -x}$$

$$= \frac{-b}{a}\left[\frac{H(x)}{H(-x)}\right]^{c_i}, \quad x > 0 ,$$

where $a,b,c_i > 0$, and H is a real-valued function satisfying

(i) H is strictly decreasing on $(-\infty, 0]$,

and (ii) H is strictly decreasing on $[0, \infty)$.

The loss to recipient i is defined by

(1.46) $$w_i[a(H(-x_i^-))^{c_i} - b(H(x_i^+))^{c_i}]$$

and the total loss to recipients will be taken to be a strictly increasing function of

(1.47) $$\Sigma\, w_i[a(H(-x_i^-))^{c_i} - b(H(x_i^+))^{c_i}]$$

where $x_i = f_i - \theta_i$ and $a,b,c_i,w_i > 0$. Extending (1.47) to account for loss to the source when $M_f \neq M_\theta$, the measure of total loss from misallocation is defined to be some strictly increasing function of

(1.48) $$\Sigma\, w_i[(H(-x_i^-))^{c_i} - b(H(x_i^+))^{c_i}] + w[a(H(x.^+))^{c} - b(H(x.^-))^{c}]$$

where $a,b,c_i,w_i > 0$.

Loss functions based on the power functions will be considered first: for $x \geq 0$ let $H(-x) = (x^-)^A$ and $H(x) = x^B$ where $A,B > 0$. Then (1.48) has the form

$$(1.49) \qquad \Sigma\, w_i[a(-x_i^-)^{Ac_i} - b(x_i^+)^{Bc_i}] + w[a(x.^+)^{Ac} - b(-x.^-)^{Bc}]$$

Proposition 1.9

Under regularity condition (1.35), if a loss function based on (1.49) is Fisher-consistent, then it is a strictly increasing function of $L_w(\underset{\sim}{f},\underset{\sim}{\theta})$ defined by (1.15).

Proof: When $\underset{\sim}{f} = \underset{\sim}{\theta}$, Fisher-consistency implies (1.49) = 0 . It is thus sufficient to show that if (1.49) > 0 for $\underset{\sim}{f} \neq \underset{\sim}{\theta}$ then $Ac_i = Bc_i = Ac = Bc = 1$ for all c_i . To see that $A = B$ and $c = c_i = \ldots = c_n$, consider $\underset{\sim}{f} - \underset{\sim}{\theta} = (x,-x,0,\ldots 0)^T$. Since (1.49) > 0 as $x \to \infty$, it follows that $c_2A \geq c_1B_1$. Considering x decreasing to 0 leads to the reverse inequality $c_2A \leq c_1B$. Thus $c_2A = c_1B$. The equality $c_1A = c_2B$ is obtained analogously. Thus $c_1 = c_2$ and $A = B$. Similarly, $c_2 = c_3 = \ldots = c_n$. To show $c = c_1$ consider $\underset{\sim}{f} - \underset{\sim}{\theta} = (x,0,\ldots 0)^T$ and proceed as above. Thus $A = B$, $c = c_1 \ldots = c_n$. Theorem 1.8 now applies to yield $Ac_i = Bc_i = Ac = Bc = 1$.

*****

Proposition 1.9 implies that no new loss functions arise by considering $H(x^+)$, $H(-x^-)$ as power functions, even in the more general context of (1.45). Fisher-consistency thus provides considerable motivation for looking at the loss function $L(\underset{\sim}{f},\underset{\sim}{\theta})$, since $L_H(\underset{\sim}{f},\underset{\sim}{\theta})$ is asymptotically $L_w(\underset{\sim}{f},\underset{\sim}{\theta})$, (1.49) is essentially $L_w(\underset{\sim}{f},\underset{\sim}{\theta})$, and the weights w_i in $L_w(\underset{\sim}{f},\underset{\sim}{\theta})$ cannot vary significantly.

Note that it is not necessarily Fisher-consistency per se that is so essential but the need to avoid pathological loss functions.

A rather different measure of loss is provided by the exponential loss functions previously introduced in (1.8), (1.11). Unlike the other loss functions considered so far, the exponential loss function accomodates significant weighting schemes. For $x > 0$ define $H(x) = -e^{-x}$, $H(-x) = e^{x}$. Then the right hand side of (1.45) equals

$$(1.50) \qquad (b/a)\, e^{-2xc_i} \qquad x > 0 .$$

The limiting behavior of (1.50) as x decreases to zero provides interpretation for the ratio b/a: for any recipient, b/a is the ratio of the benefit from an infinitesimally small overallocation to the loss arising from an underallocation of the same absolute magnitude. We shall assume for simplicity that $b/a = 1$. Then (1.46) has the form

$$(1.51) \qquad w_i\, e^{-c_i x_i}$$

In accordance with (1.48), the total loss from errors in allocation is taken to be a strictly increasing function of

$$(1.52) \qquad \Sigma\, w_i\, e^{-c_i x_i} + w\, e^{cx.} .$$

Exponential loss functions have been used as measures of inequity by several authors. As mentioned in §1.2 the Stanford Research Institute (1974b) used exponential loss to assess the inequity of errors in GRS allocations; J. Ferreira (1978) uses exponential loss functions for determining an equitable distribution of automobile insurance rates among different risk classes. Neither of these uses of exponential loss functions explicitly confronted the benefit-cost problem of deciding how much should be spent on better data to reduce the level of inequity. To use exponential loss in a benefit-cost study it seems appropriate to take logs in (1.52). The Stanford Research Institute did this, obtaining essentially (1.11') to measure total loss from inequity. Taking logs in (1.52) and translating to obtain zero loss when $\underset{\sim}{f} = \underset{\sim}{\theta}$ yields the loss function

$$(1.53)\qquad L_E(\underset{\sim}{f},\underset{\sim}{\theta}) = \log[\Sigma w_i e^{-c_i x_i} + w e^{cx.}] - \log(w + \Sigma w_i)$$

where $x_i = f_i - \theta_i$, $x. = M_f - M_\theta$.

Proposition 1.10 Under the regularity condition (1.35), $L_E(\underset{\sim}{f},\underset{\sim}{\theta})$ is Fisher-consistent if and only if

$$(1.54)\qquad wc = w_1 c_1 = \dots = w_n c_n = \lambda > 0 \ .$$

Proof: It suffices to show that (1.54) is necessary and sufficient for (1.53) to attain its unique minimum at $\underset{\sim}{f} = \underset{\sim}{\theta}$. If $\underset{\sim}{f} - \underset{\sim}{\theta} = (x,0,\dots 0)^T$ then (1.53) equals $\Psi(x)$, where $\Psi(x) = w_1 e^{-c_1 x} + w e^{cx.}$. Fisher-consistency implies $0 = \Psi'(0) = -w_1 c_1 + wc$. Thus $wc = w_1 c_1$ is necessary for Fisher-consistency. The remainder of (1.54) is similarly proved.

Now we show (1.54) implies Fisher-consistency. Without loss of generality set $\lambda = 1$. It is sufficient to show that the unique minimum of

$$G(\underset{\sim}{x}) = \Sigma c_i^{-1} e^{-c_i x_i} + c^{-1} e^{c\Sigma x_i}$$

occurs at $\underset{\sim}{x} = \underset{\sim}{0}$. The ith element of the gradient vector of first partial derivatives of G is $-e^{-c_i x_i} + e^{c\Sigma x_i}$, so the gradient equals $\underset{\sim}{0}$ only at $\underset{\sim}{x} = \underset{\sim}{0}$. To show G attains its unique minimum at $\underset{\sim}{x} = \underset{\sim}{0}$ it is sufficient to show that the Hessian matrix Γ is positive definite at $\underset{\sim}{x} = \underset{\sim}{0}$. The ij^{th} entry of Γ is the second partial derivative $\partial^2 G/(\partial x_i \partial x_j)$ at $\underset{\sim}{0}$: $c + c_i$ if $i = j$ and c if $i \neq j$. To see that Γ is positive definite consider mutually independent random variables $U, Z_1, \ldots, Z_n$ such that $\text{Var}(U) = c$, $\text{Var}(Z_i) = c_i$. Let $X_i = Z_i + U$. Then Γ equals the covariance matrix of $X_1, \ldots, X_n$, which is positive definite because the Z_i's are independent.

*****

As we shall see in chapter 2 it is generally necessary to use approximations when working with expected loss. Unfortunately the approximations for expected loss under (1.53) are poor approximations, being essentially expressions of expected squared-error loss. In practice then, we cannot use (1.53) for benefit-cost analysis, but must use either a poor approximation or another loss criterion entirely.

In the sequel we will use the loss function $L(\underset{\sim}{f}, \underset{\sim}{\theta})$ given by (1.16) or (1.26). Technical reasons (such as Fisher-consistency) for using this loss function were given in §1.6 above. There are other, more apparent, practical reasons as well. The loss function $L(\underset{\sim}{f}, \underset{\sim}{\theta})$, which in a fixed-pie allocation formula equals $(a-b)/2$ times the sum of the absolute errors in allocation $\Sigma|f_i - \theta_i|$, is easy to interpret and understand. For example, for a fixed-pie allocation formula $L(\underset{\sim}{f}, \underset{\sim}{\theta})$ is a constant times the number of dollars that would have to be redistributed to achieve perfectly accurate allocations. A heuristic interpretation of $L(\underset{\sim}{f}, \underset{\sim}{\theta})$ as a measure of social loss has also been provided in §1.3 where the constants a and b were interpreted as interest rates.

We will not use the loss function $L(\underset{\sim}{f}, \underset{\sim}{\theta})$ without restriction. As mentioned in §1.2 in discussion of Rawls's and Harsanyi's views of social justice, undiscriminating use of an "average loss" criterion such as $L(\underset{\sim}{f},\underset{\sim}{\theta})$ can lead to drastic errors for certain subgroups of the population. The political consequences could be severe. We will use $L(\underset{\sim}{f},\underset{\sim}{\theta})$ subject to the restriction that individual errors must fall within tolerance limits. For example, we might require that no underallocation shall exceed sixty percent of the optimal allocation. Use of restrictions such as this relates closely to the "restricted Bayes criterion" (Lehmann 1959) and other notions in the statistical literature of "limiting the risk" (Efron and Morris 1971; Fay and Herriot 1979). In subsequent applications of the loss function we will assume that the restrictions are satisfied.

Chapter 2

The Delta Method

This chapter is highly technical and is included to show that the approximations for moments of errors in data (chapter 3) and in allocations (chapters 4 and 5) can be developed rigorously. The reader is encouraged to skip this chapter for now and return to it if motivated by a desire to verify approximations developed in later chapters.

§ 2.0 Outline

The delta method is a procedure for calculating asymptotic approximations to random variables and their distributions, means, variances, and covariances. Standard theory (as in Cramer 1945, pp. 352 ff., 366 ff.; and Bishop, Fienberg, and Holland 1975, pp. 486 ff.) applies when the random variables can be thought of as being generated by some kind of repetitive sampling mechanism. The data series under our consideration typically cannot be thought of as being generated in this way and a more general stochastic model is needed. In particular, the stochastic model we use allows for biases and weaker orders of magnitude for the remainder terms.

Extensive notation (§2.1) is needed to provide a rigorous description of the delta method (§2.2). Because the allocation formulas are not "smooth", demonstration that the delta method is applicable for assessing moments of errors in GRS requires some effort (§2.3). The delta method also applies to yield approximations to expected loss (§2.4) under the loss functions considered in chapter 1.

§ 2.1 Notation

For vectors $\underset{\sim}{x}$ and $\underset{\sim}{y}$ in R^k and real-valued functions $f_j: R^k \to R^1$, $j=1,\ldots,q$, possessing first partial derivatives $\frac{\partial f_j}{\partial x_i}(\underset{\sim}{x})$ define

$$||\underset{\sim}{x}|| = (\sum_i x_i^2)^{1/2}$$

$$\underset{\sim}{x} \cdot \underset{\sim}{y} = \sum_i x_i y_i$$

$$\nabla f_j(\underset{\sim}{x}) = (\frac{\partial f_j}{\partial x_1}(\underset{\sim}{x}),\ldots,\frac{\partial f_j}{\partial x_k}(\underset{\sim}{x}))^T$$

and $\nabla \underset{\sim}{f}(\underset{\sim}{x}) = (\nabla f_1(\underset{\sim}{x}), \ldots, \nabla f_q(\underset{\sim}{x}))$ ($\nabla \underset{\sim}{f}$ is a $k \times q$ matrix),

where the symbol ^{T} denotes the transpose of a matrix. For a real-valued function $g: R^k \to R^1$ possessing first and second derivatives denote the Hessian matrix by $\nabla^2 g(\underset{\sim}{x}) = \nabla\nabla g(\underset{\sim}{x})$, so that the ij^{th} entry of $\nabla^2 g(\underset{\sim}{x})$ is $\partial^2 g(\underset{\sim}{x})/\partial x_i \partial x_j$.

Let $\{\xi_N\}$ be a monotone sequence of positive real numbers. Then for the sequence $\{\psi_N\}$ the following notation is used:

$$\psi_N = o(\xi_N) \quad \text{if} \quad \lim_{N \to \infty} \psi_N/\xi_N = 0$$

and

$$\psi_N = O(\xi_N) \quad \text{if} \quad \overline{\lim_{N \to \infty}} |\psi_N/\xi_N| < \infty .$$

The notation is generalized to include vectors $\underset{\sim}{\psi}_N$ so that

$$\underset{\sim}{\psi}_N = o(\xi_N) \quad \text{if} \quad \lim_{N \to \infty} ||\underset{\sim}{\psi}_N||/\xi_N = 0$$

and

$$\underset{\sim}{\psi}_N = O(\xi_N) \quad \text{if} \quad \overline{\lim_{N \to \infty}} ||\underset{\sim}{\psi}_N||/\xi_N < \infty .$$

The notation extends to random sequences $\{X_N\}$ of real-valued random variables by defining

$$X_N = o_p(\xi_N) \quad \text{if} \quad \text{for all } \varepsilon > 0 \quad \lim_{N \to \infty} \text{Prob}(|X_N|/\xi_N > \varepsilon) = 0$$

and

$$X_N = O_p(\xi_N) \quad \text{if} \quad \text{for all } \eta > 0 \text{ there exists } \varepsilon > 0 \text{ such that } \lim_{N \to \infty} \text{Prob}(|X_N|/\xi_N > \varepsilon) < \eta .$$

The notation extends further to sequences of random vectors $\{\underset{\sim}{X}_N\}$ by defining

$$\underset{\sim}{X}_N = o_p(\xi_N) \quad \text{if} \quad ||\underset{\sim}{X}_N|| = o_p(\xi_N)$$

and $\quad \mathbf{X}_N = O_p(\xi_N)$ if $\|\mathbf{X}_N\| = O_p(\xi_N)$.

Finally, a sequence $\{W_N\}$ of real matrices is $o(\xi_N)$ or $O(\xi_N)$ if $\|W_N\|$ is $o(\xi_N)$ or $O(\xi_N)$, where $\|W_N\|^2 = \sum_i\sum_j w^2_{N,i,j}$ for $w_{N,i,j}$ the ij^{th} entry of W_N .

§ 2.2 Description of the Delta Method

The concept of a sequence of random vectors being asymptotically normal will be required, although not in its most general formulation. Let $\{\boldsymbol{\mu}_N\}$ and $\{\boldsymbol{\Sigma}_N\}$ be sequences of k×1 vectors and k×k matrices respectively, satisfying

$$(2.1) \qquad \boldsymbol{\mu}_N = \boldsymbol{\mu} + \mathbf{b}N^{-1/2} + o(N^{-1/2}) \quad \text{and} \quad \boldsymbol{\Sigma}_N = \boldsymbol{\Sigma}N^{-1} + o(N^{-1}) .$$

We say that $\mathbf{X}_N$ is AMVN$(\boldsymbol{\mu}_N, \boldsymbol{\Sigma}_N)$ provided (2.1) holds and for all $\boldsymbol{\lambda}$ such that $\boldsymbol{\lambda}^T\boldsymbol{\Sigma}\boldsymbol{\lambda} > 0$,

$$(2.2) \qquad \lim_{N \to \infty} \text{Prob}\left(\frac{\boldsymbol{\lambda}^T(\mathbf{X}_N - \boldsymbol{\mu}_N)}{(\boldsymbol{\lambda}^T\boldsymbol{\Sigma}_N\boldsymbol{\lambda})^{1/2}} \leq x\right) = \Phi(x) .$$

Note that if $\mathbf{X}_N$ is AMVN$(\boldsymbol{\mu}_N, \boldsymbol{\Sigma}_N)$ then $\mathbf{X}_N$ is AMVN$(\boldsymbol{\mu}_N, \boldsymbol{\Sigma}N^{-1})$. To see this notice

$$\frac{\boldsymbol{\lambda}^T(\mathbf{X}_N - \boldsymbol{\mu}_N)}{(\boldsymbol{\lambda}^T\boldsymbol{\Sigma}N^{-1}\boldsymbol{\lambda})^{-1/2}} = [1 + o(1)]\frac{\boldsymbol{\lambda}^T(\mathbf{X}_N - \boldsymbol{\mu}_N)}{(\boldsymbol{\lambda}^T\boldsymbol{\Sigma}_N\boldsymbol{\lambda})^{1/2}}$$

and realize that the asymptotic distribution of the left-hand side does not depend on the term $o(1)$. This comment points out that at least in theory it is no better to use $\boldsymbol{\Sigma}_N$ than $\boldsymbol{\Sigma}N^{-1}$ in the asymptotic calculations. Furthermore, it is often easier to evaluate $\boldsymbol{\Sigma}$.

Sometimes the sequence $\{\underset{\sim}{X}_N\}$ is of the form $\{\sum_{i=1}^{N} \underset{\sim}{Y}_i/N\}$, where $\underset{\sim}{Y}_i$ are independent and identically distributed with mean $\underset{\sim}{\mu}$ and variance-covariance matrix Σ, so that the central limit theorem applies (Anderson 1958, p.74). In the applications before us, it will be appropriate to assume that $\underset{\sim}{X}_N$ is asymptotically multivariate normal, partly because of the central limit theorem. However, the situations will be complex and the assumption must finally be based on empirical evidence and intuition. The index N will not always be interpretable as a sample size. Sometimes it will refer to the cost of a data program or perhaps it will have a somewhat vague interpretation as the "stage of development" of a data program.

<u>Theorem 2.1</u> (The Delta Method)
Assume (2.1) holds and let $\underset{\sim}{X}_N$ satisfy

(2.3) $E\underset{\sim}{X}_N = \underset{\sim}{\mu}_N$

(2.4) $\text{Var}\ \underset{\sim}{X}_N = \Sigma_N$

(2.5) $\underset{\sim}{X}_N$ is $\text{AMVN}(\underset{\sim}{\mu}_N, \Sigma_N)$ and

(2.6) all fourth central moments of $\underset{\sim}{X}_N$ are $o(N^{-1})$.

Consider $\underset{\sim}{h}(\underset{\sim}{x}) = (h_1(\underset{\sim}{x}), \ldots, h_q(\underset{\sim}{x}))^T$ and assume $h_1, \ldots, h_q\colon R^k \to R^1$ are bounded functions with continuous and bounded first and second derivatives. Then

(2.7) $\underset{\sim}{h}(\underset{\sim}{X}_N) = \underset{\sim}{h}(\underset{\sim}{\mu}) + \nabla\underset{\sim}{h}(\underset{\sim}{\mu})^T(\underset{\sim}{X}_N - \underset{\sim}{\mu}) + o_p(N^{-1/2})$

(2.8) $\underset{\sim}{h}(\underset{\sim}{X}_N)$ is $\text{AMVN}(\ \underset{\sim}{h}(\underset{\sim}{\mu}) + \nabla\underset{\sim}{h}(\underset{\sim}{\mu})^T\underset{\sim}{b}N^{-1/2}\ ,\ \nabla\underset{\sim}{h}(\underset{\sim}{\mu})^T\Sigma\nabla\underset{\sim}{h}(\underset{\sim}{\mu})N^{-1}\ )$

(2.9) $E\underset{\sim}{h}(\underset{\sim}{X}_N) = \underset{\sim}{h}(\underset{\sim}{\mu}) + \nabla\underset{\sim}{h}(\underset{\sim}{\mu})^T\underset{\sim}{b}N^{-1/2} + o(N^{-1/2})$

and

$$(2.10) \qquad \operatorname{Var} \underset{\sim}{h}(\underset{\sim}{X}_N) = \nabla \underset{\sim}{h}(\underset{\sim}{\mu})^T \Sigma \nabla \underset{\sim}{h}(\underset{\sim}{\mu}) N^{-1} + o(N^{-1}) .$$

<u>Proof</u>: We use ε_{ij} to denote bounded random variables. First consider q=1. Expanding $\underset{\sim}{h}(\underset{\sim}{X}_N)$ in Taylor series and taking advantage of the bounded second derivatives of h we obtain

$$(2.11) \qquad h(\underset{\sim}{X}_N) = h(\underset{\sim}{\mu}) + \nabla h(\underset{\sim}{\mu})^T(\underset{\sim}{X}_N-\underset{\sim}{\mu}) + \sum_{ij}\sum \varepsilon_{ij}(X_{N,i}-\mu_i)(X_{N,j}-\mu_j).$$

Since $\underset{\sim}{X}_N-\underset{\sim}{\mu} = O_p(N^{-1/2})$, it follows that we obtain a slight improvement on (2.7):

$$(2.12) \qquad h(\underset{\sim}{X}_N) = h(\underset{\sim}{\mu}) + \nabla h(\underset{\sim}{\mu})^T(\underset{\sim}{X}_N-\underset{\sim}{\mu}) + O_p(N^{-1}) .$$

Furthermore, since $\underset{\sim}{X}_N-\underset{\sim}{\mu} = O_p(N^{-1/2})$, the asymptotic distribution of $h(\underset{\sim}{X}_N)$ is the same as that of $h(\underset{\sim}{\mu}) + (\underset{\sim}{X}_N-\underset{\sim}{\mu})^T\nabla h(\underset{\sim}{\mu})$, and thus (2.8) holds. To find the asymptotic expected value of $h(\underset{\sim}{X}_N)$ compute from (2.11):

$$Eh(\underset{\sim}{X}_N) = h(\underset{\sim}{\mu}) + E\nabla h(\underset{\sim}{\mu})^T(\underset{\sim}{X}_N-\underset{\sim}{\mu}) + E(\sum_{ij}\sum \varepsilon_{ij}(X_{N,i}-\mu_i)(X_{N,j}-\mu_j) .$$

Letting σ_{ij} denote the ij^{th} element of Σ and b_i denote the ith element of $\underset{\sim}{b}$ we have

$$\begin{aligned} E(X_{N,i}-\mu_i)(X_{N,j}-\mu_j) &= E[(X_{N,i}-\mu_{N,i}+b_iN^{-1/2}+o(N^{-1/2})) \\ &\qquad (X_{N,j}-\mu_{N,j}+b_jN^{-1/2}+o(N^{-1/2}))] \\ &= \sigma_{ij}N^{-1} + b_ib_jN^{-1} + o(N^{-1}) \\ &= O(N^{-1}) \end{aligned}$$

and consequently

$$(2.13)\qquad Eh(\underset{\sim}{X}_N) = h(\underset{\sim}{\mu}) + \nabla h(\underset{\sim}{\mu})^T \underset{\sim}{b} N^{-1/2} + o(N^{-1/2}) .$$

To find the asymptotic variance of $h(\underset{\sim}{X}_N)$ use (2.13) and (2.11):

$$\begin{aligned} \text{Var } h(\underset{\sim}{X}_N) &= E[\ h(\underset{\sim}{\mu}) + \nabla h(\underset{\sim}{\mu})^T(\underset{\sim}{X}_N - \underset{\sim}{\mu}) + \Sigma\Sigma\ \varepsilon_{ij}(X_{N,i}-\mu_i)(X_{N,j}-\mu_j) \\ &\qquad - h(\underset{\sim}{\mu}) - \nabla h(\underset{\sim}{\mu})^T \underset{\sim}{b} N^{-1/2} + o(N^{-1/2})\ ]^2 \\ &= E[\ \nabla h(\underset{\sim}{\mu})^T(\underset{\sim}{X}_N - \underset{\sim}{\mu}_N) + \Sigma\Sigma\ \varepsilon_{ij}(X_{N,i}-\mu_i)(X_{N,j}-\mu_j) \\ &\qquad + o(N^{-1/2})\ ]^2 \\ &= \nabla h(\underset{\sim}{\mu})^T\ \Sigma_N\ \nabla h(\underset{\sim}{\mu}) + 2\ E[\ \nabla h(\underset{\sim}{\mu})^T \cdot (\underset{\sim}{X}_N - \underset{\sim}{\mu}_N) \cdot \\ &\qquad \Sigma\Sigma\ \varepsilon_{ij}(X_{N,i}-\mu_i)(X_{N,j}-\mu_j)\ ] \\ &\qquad + E(\Sigma\Sigma\ \varepsilon_{ij}(X_{N,i}-\mu_i)(X_{N,j}-\mu_j)\ )^2 + o(N^{-1}) \\ &= \nabla h(\underset{\sim}{\mu})^T\ \Sigma\ \nabla h(\underset{\sim}{\mu})\ N^{-1} + o(N^{-1}) \end{aligned}$$

which is (2.10).

Now consider $q > 1$. Expressions (2.7) and (2.9) are proved by componentwise applications of the theorem for $q=1$. Now let $\underset{\sim}{A}$ be an arbitrary $q \times 1$ vector. Applying the theorem for $q=1$ we observe that $\underset{\sim}{A}^T h(\underset{\sim}{X}_N)$ is AMVN($\underset{\sim}{A}^T(\underset{\sim}{h}(\underset{\sim}{\mu}) + \nabla \underset{\sim}{h}(\underset{\sim}{\mu})^T \underset{\sim}{b} N^{-1/2})$, $\underset{\sim}{A}^T \nabla \underset{\sim}{h}(\underset{\sim}{\mu})^T \Sigma_N \nabla \underset{\sim}{h}(\underset{\sim}{\mu}) \underset{\sim}{A}$) . Since $\underset{\sim}{A}$ was arbitrary, (2.8) is established for $q \geq 1$. Similarly, since (2.10) holds for $q=1$ we have

$$\text{Var } \underset{\sim}{A}^T \underset{\sim}{h}(\underset{\sim}{X}_N) = \underset{\sim}{A}^T\ \nabla \underset{\sim}{h}(\underset{\sim}{\mu})^T\ \Sigma_N\ \nabla \underset{\sim}{h}(\underset{\sim}{\mu})\ \underset{\sim}{A} + o(N^{-1/2}) .$$

The arbitrariness of $\underset{\sim}{A}$ establishes (2.10) for $q \geq 1$.

*****

Notice that the strong assumptions made in the theorem ensure that formal manipulation of (2.7) to obtain (2.9) and (2.10) is justified. Specifically, (2.9) and (2.10) can be derived from (2.7) by taking moments of the main terms and replacing the $o_p(N^{-1/2})$ by $o(N^{-1/2})$ or $o(N^{-1})$. This method of evaluating (2.9) and (2.10) is often more convenient than direct calculation.

The following result is standard and easy to derive. We state it explicitly for future reference.

<u>Corollary 2.2</u>

Let $\{\underset{\sim}{X}_N\}$ satisfy the hypotheses of theorem 2.1, where $\updownarrow$ is a diagonal matrix and $X_{N,i} \geq 0$, $X_{N\cdot} = \sum_j X_{N,j} > \delta > 0$ (with probability one). Then for the function $\underset{\sim}{h}=(h_1,\ldots,h_k)$ defined by

$$h_i(\underset{\sim}{x}) = \begin{cases} x_i/\Sigma\, x_j & \Sigma\, x_j \neq 0 \\ 0 & \Sigma\, x_j = 0 \end{cases}$$

the covariance between $h_i(\underset{\sim}{X}_N)$ and $h_j(\underset{\sim}{X}_N)$ is

$$(2.14)\qquad \mu_\cdot^{-2}\left(\frac{-\mu_j}{\mu_\cdot}\,\sigma_i^2 - \frac{\mu_i}{\mu_\cdot}\,\sigma_j^2 + \frac{\mu_i}{\mu_\cdot}\frac{\mu_j}{\mu_\cdot}\Big(\sum_k \sigma_k^2\Big)\right) + o(N^{-1}) \qquad i \neq j$$

and

$$(2.15)\qquad \mu_\cdot^{-2}\left[1 - \left(\frac{2\mu_i}{\mu_\cdot}\right)\sigma_i^2 + \frac{\mu_i^2}{\mu_\cdot^2}\Big(\sum_k \sigma_k^2\Big)\right] + o(N^{-1}) \qquad i = j,$$

where σ_i^2 is the ith diagonal entry of $\updownarrow$ and $\mu_\cdot = \Sigma\, \mu_k$.

<u>Proof</u>: First we verify that theorem 2.1 applies. For $\Sigma\, X_\ell > \delta$ the function h_i and its first two derivatives are sums of terms of the

form $(X_j)^r(\Sigma X_\ell)^{-s}$ for $r=0,1$ and $s=1,2,3$. Certainly $X_{N.} > \delta$ so that h and its first two derivatives are bounded and continuous functions. Thus theorem 2.1 applies. Noting that the ij^{th} entry of $\nabla \underset{\sim}{h}(\underset{\sim}{\mu})$ is

$$(\mu_. - \mu_i)/\mu_.^2 \qquad i=j$$

and

$$- \mu_j/\mu_.^2 \qquad i\neq j$$

we can easily show that the ij^{th} entry of $\nabla \underset{\sim}{h}(\underset{\sim}{\mu})^T \, \underset{\sim}{\Sigma} \, \nabla \underset{\sim}{h}(\underset{\sim}{\mu})$ is given by the leading terms of (2.14) and (2.15) for $i\neq j$ and $i=j$ respectively.

*****

§ 2.3 Applicability of the Delta Method to GRS

In applying the delta method we let $\underset{\sim}{X}_N$ denote the data input for the GRS allocation formulas. For some of the components of $\underset{\sim}{X}_N$ we may interpret N as a sample size (e.g. state federal individual income tax liabilities, discussed in §3.10). For other data, such as state population discussed in §3.2, N has a vague interpretation, such as "stage of development" of the data program or cost of the data program. We may think of N as being asymptotic because the relative standard errors and relative absolute biases are small, typically less than 0.10. Formula (2.1) holds because the relative biases and relative standard errors are empirically of the same order of magnitude. Although the original allocation formulas do not mention relative errors, analysis of the formulas by the delta method leads us to focus on the relative errors. Their roles will be explained in greater detail later in this section.

The assumption (2.5) of asymptotic normality is motivated by central limit theory considerations, viz. the vector $\underset{\sim}{X}_N-\underset{\sim}{\mu}_N$ of random errors in the data is the sum of many small random errors. For example, errors in the estimates of state population updates (see §3.2) arise from errors in vital statistics, elementary school enrollments data, counts of the number of federal individual income tax returns filed, passenger automobile registrations data, data on the work force, Medicare data, and other data, as well errors arising from the methodology used (Census 1974b, No. 520).

To analyze the state-level allocations in GRS let h denote the allocation to state i. The first step in computing h is to calculate the "House formula amount" f and the "Senate formula amount" g (described explicitly in §4.1 below). The allocation to state i will be determined by whichever of the two formulas f or g give it a larger allocation, say d , given by

$$(2.16) \quad d(\underset{\sim}{x}) = I_A(\underset{\sim}{x})\, f(\underset{\sim}{x}) + I_B(\underset{\sim}{x})\, g(\underset{\sim}{x})$$

where A and B are disjoint sets whose union is R^k and I_A and I_B are the indicator functions for the sets A and B. For a typical state $A = \{\underset{\sim}{x}: f(\underset{\sim}{x}) \geq g(\underset{\sim}{x})\}$ and $B = \{\underset{\sim}{x}: f(\underset{\sim}{x}) < g(\underset{\sim}{x})\}$. The allocation $h(\underset{\sim}{x})$ to state i is proportional to $d(\underset{\sim}{x})$ divided by the sum of the values of $d(\underset{\sim}{x})$ for all states.

The functions f and g are sums of terms like Q_j given by

$$(2.17) \quad Q_j = X_j/\sum_\ell X_\ell \ .$$

With probability one, $X_j \geq 0$ and $\Sigma X_\ell > 0$ so Q_j and hence f and g are all bounded. For example, if $\underset{\sim}{X}$ refers to the population of states in 1970 then certainly $\Sigma X_\ell > 180{,}000{,}000$ and $0.001 < Q_j < 0.110$. Since the first two derivatives of Q_j are sums of terms of the form $(X_j)^r(\Sigma X_\ell)^{-s}$ for r=0,1 and s=1,2,3, all first and second derivatives of f and g are bounded and continuous (with probability one).

In applying the delta method we will substitute for d either f or g, depending on which formula was used to calculate the state's allocation. The motivation is that f and g satisfy the hypotheses of theorem 2.1 while d does not. Fortunately, the error introduced by this substitution will be small because f equals g on the boundary of A and B and because f and g react the same way to small changes in $\underset{\sim}{x}$ near the boundary. For example, both f and g are increasing functions of population and decreasing functions of per capita income. For $\underset{\sim}{x}$ far from the boundary the substitution of f or g for d will be correct with high probability. Application of theorem 2.1 to situations such as this will be referred to as the "modified" delta method.

The substate allocation formulas (presented in chapter 5) are far more complicated than the state allocation formulas but are composed of the same kinds of components, i.e. (2.16) and (2.17). To apply the delta method to the substate allocation formulas, substitutions similar to that above (i.e. f or g for d) will be made. For the reasons given above, the error introduced by these substitutions is expected to be small.

In studying state-level data elements in chapter 3 attention will be focused on terms of the form

$$(2.18)\qquad \frac{X_j - \mu_j}{\mu_j} - \frac{X_. - \mu_.}{\mu_.} \ .$$

The reason for studying these deviations of state relative errors about the national average is the central role of terms like Q_j in the allocation formulas. The deviation of Q_j from its parameter value is approximated (using (2.7)) by

$$\frac{\mu_j}{\mu_.}\left(\frac{X_j - \mu_j}{\mu_j} - \frac{X_. - \mu_.}{\mu_.}\right) \ .$$

To estimate biases and variances in states' allocations it will be necessary to estimate the means and variances of (2.18). This will be done in chapter 3.

§ 2.4 Application of the Delta Method to Calculating Expected Loss

The expected values of the loss functions $L(\underset{\sim}{h},\underset{\sim}{\theta})$ and $L_E(\underset{\sim}{h},\underset{\sim}{\theta})$ considered in chapter 1 (see (1.26) and (1.53)) are difficult to compute exactly. The following proposition extends theorem 2.1 in a special case, covering (1.53) but not (1.26).

Proposition 2.3

Let $\{\underset{\sim}{Y}_N\}$ be a sequence of random vectors satisfying

(i) $E\underset{\sim}{Y}_N = \underset{\sim}{U} + \underset{\sim}{M}N^{-1/2} + o(N^{-1/2})$

(ii) $\operatorname{Var} \underset{\sim}{Y}_N = WN^{-1} + o(N^{-1})$

(iii) $\underset{\sim}{Y}_N$ is $\operatorname{AMVN}(\underset{\sim}{U}+\underset{\sim}{M}N^{-1/2}, WN^{-1})$ and

(iv) all third moments of $\underset{\sim}{Y}_N$ are $o(N^{-1})$

and let G be a real-valued function possessing bounded and continuous third derivatives such that $\nabla G(\underset{\sim}{U}) = 0$. Then

$$(2.19) \quad EG(\underset{\sim}{Y}_N) = G(\underset{\sim}{U}) + \frac{1}{2N} \Sigma\Sigma(w_{ij} + m_i m_j)\nabla_{ij} + o(N^{-1})$$

where w_{ij} is the ij^{th} element of W, m_i is the ith element of $\underset{\sim}{M}$, and ∇_{ij} is the ij^{th} element of $\nabla^2 G(\underset{\sim}{U})$.

Proof: By Taylor expansion about $\underset{\sim}{U}$ we have

$$G(\mathbf{Y}_N) \;=\; G(\mathbf{U}) + (\mathbf{Y}_N-\mathbf{U})^T\, \nabla G(\mathbf{U}) \;+\; \tfrac{1}{2}(\mathbf{Y}_N-\mathbf{U})^T\, \nabla^2 G(\mathbf{U})\, (\mathbf{Y}_N-\mathbf{U})$$

$$+ \sum_{ijk}\sum\sum \varepsilon_{ijk}(Y_{N,i} - u_i)(Y_{N,j} - u_j)(Y_{N,k} - u_k)$$

where ε_{ijk} denotes a bounded random variable. Since $\nabla G(\mathbf{U})=0$, we have

$$\text{(2.20)} \qquad EG(\mathbf{Y}_N) = G(\mathbf{U}) + E(\mathbf{Y}_N-\mathbf{U})^T\, \nabla^2 G(\mathbf{U})\, (\mathbf{Y}_N-\mathbf{U}) + o(N^{-1}),$$

and (2.19) is immediate.

*****

To apply the theorem let $\mathbf{X}_N$ and $\mathbf{h}$ satisfy the hypotheses of theorem 2.1 and let G be any smooth Fisher-consistent loss function depending on $\mathbf{Y}_N=\mathbf{h}(\mathbf{X}_N)-\mathbf{h}(\boldsymbol{\mu})$. Set $\mathbf{U}=\mathbf{0}$, $\mathbf{M}=\nabla\mathbf{h}(\boldsymbol{\mu})^T\mathbf{b}$, and $W=\nabla\mathbf{h}(\boldsymbol{\mu})^T\Sigma\nabla\mathbf{h}(\boldsymbol{\mu})$. Then theorem 2.1 implies (i)-(iv), and so by the preceding proposition, $EG(\mathbf{Y}_N)$ equals

$$\text{(2.21)} \qquad \frac{1}{2N} \sum\sum \nabla_{ij}(w_{ij} + m_i m_j) \;+\; o(N^{-1})$$

or equivalently

$$\text{(2.22)} \qquad \tfrac{1}{2}(E\mathbf{Y}_N)^T\, \nabla^2 G(\mathbf{0})\, \mathbf{Y}_N \;+\; o(N^{-1}) \;.$$

Thus when the allocations are near the optimum, the risk under G can be approximated by a non-negative quadratic form. In particular, the risk under $L_E(\mathbf{h},\boldsymbol{\theta})$ will be a (positive) linear combination of the elements of $W + \mathbf{M}\mathbf{M}^T$.

The preceding proposition fails to apply to (1.26) because the first derivative does not everywhere exist. To calculate expected loss under the loss function $L(\mathbf{h},\boldsymbol{\theta})$ given by (1.26) we need only approximate $E|h(\mathbf{X}_N) - h(\boldsymbol{\mu})|$ for scalar functions h. The following proposition shows how to proceed, but notice that the remainder term is $o(N^{-1/2})$ rather than $o(N^{-1})$ as in proposition 2.3.

Proposition 2.4

Assume that $\{\underset{\sim}{X}_N\}$ is a sequence of random variables and h is a scalar function such that the hypotheses of theorem 2.1 are satisfied and $\nabla h(\underset{\sim}{\mu})^T \ddagger \nabla h(\underset{\sim}{\mu}) > 0$. Then

$$(2.23) \quad E|h(\underset{\sim}{X}_N)-h(\underset{\sim}{\mu})| = B[1-2\Phi(-B/S)] + 2S\,\phi(B/S) + o(N^{-1/2})$$

where $B = \nabla h(\underset{\sim}{\mu})^T \underset{\sim}{b} N^{-1/2}$ and $S = [\ \nabla h(\underset{\sim}{\mu})^T \ddagger \nabla h(\underset{\sim}{\mu}) N^{-1}\ ]^{1/2}$. (Notice that B and S depend on N but B/S does not.)

Proof: Notice

$$E|h(\underset{\sim}{X}_N)-h(\underset{\sim}{\mu})| = S\,E\left|\frac{h(\underset{\sim}{X}_N)-h(\underset{\sim}{\mu})}{S}\right| = S\int |x|\,dF_N(x)$$

where $F_N(\ )$ is the cumulative distribution function of $[h(\underset{\sim}{X}_N)-h(\underset{\sim}{\mu})]/S$. If we can show

$$(2.24) \quad \int |x|\,dF_N(x) = \int |x|\,d\Phi(x-B/S) + o(1)$$

then we are done, since $S\cdot o(1)=o(N^{-1/2})$, and (integrate by parts)

$$(2.25) \quad \int |x|\,d\Phi(x-B/S) = \frac{B}{S}[1 - 2\Phi(-B/S)] + 2\phi(B/S)\ .$$

Theorem 2.1 implies

$$\int (x^2+1)\,dF_N(x) = \int (x^2+1)\,d\Phi(x-B/S) + o(1)$$

and $|x| < x^2+1$, so by a generalization of the Lebesgue Convergence Theorem (Royden 1968, p.89) we have (2.24).

*****

The approximation given by proposition 2.3 for the expectation of $L_E(\underset{\sim}{h},\underset{\sim}{\theta})$ is a local one. When the distribution of $\underset{\sim}{h}-\underset{\sim}{\theta}$ is sufficiently concentrated around $\underset{\sim}{0}$ this local approximation will be good. For the values of N applicable in practice (that is, for the large values of $h_i-\theta_i$ likely to occur) this local quadratic approximation to the logarithm of an exponential function will be toc large because the distribution of $\underset{\sim}{h}-\underset{\sim}{\theta}$ is not sufficiently concentrated at zero.

To see this in a simple case consider $n=2$, $w_1=w_2=w=c_1=c_2=c=1$, and $h_1-\theta_1 = -(h_2-\theta_2) = x$, and notice

$$L_E(\underset{\sim}{h},\underset{\sim}{\theta}) = \log[(e^x + e^{-x} + 1)/3] .$$

From elementary calculus it follows that

$$\lim_{x \to \infty} \frac{L_E(\underset{\sim}{h},\underset{\sim}{\theta})}{x} = 1,$$

and therefore a quadratic approximation must overstate the loss for large x.

The approximation for the expectation of the loss function $L(\underset{\sim}{h},\underset{\sim}{\theta})$ mainly involves approximating the distribution of $\underset{\sim}{h}$ by a multivariate normal distribution. Because the family of multivariate normal distributions is closed under location and scale transformations, the approximation is not only local. That is, the approximation does not require $\underset{\sim}{h}-\underset{\sim}{\theta}$ to be close to zero with high probability. Rather, it requires the distribution of $\underset{\sim}{h}-\underset{\sim}{\theta}$ to be approximately multivariate normal. This approximation is believed satisfactory for the relevant values of N (i.e. the actual distribution of $\underset{\sim}{h}-\underset{\sim}{\theta}$).

Chapter 3 Data Used in General Revenue Sharing

§ 3.1 Introduction

The GRS allocation formulas involve a large number of data elements (DE's). This chapter develops error models for each of the data elements. These error models will be used in chapter 6 to calculate the expected losses in GRS allocations, but the models are also of independent interest because the methodology extends naturally to other kinds of data as well. Construction of error models is a difficult undertaking because the mechanisms giving rise to errors in data are complicated and not generally understood. Of these mechanisms random sampling is typically the simplest and often the least important. To derive the error model for a data element we proceed in five steps: (1) notation; (2) definition and data sources; (3) data quality: evidence; (4) data quality: assumptions and conclusions; and (5) relevant distributions and approximations.

Each of the data series used in GRS is discussed in a separate section; each section divides into subsections corresponding to the five steps mentioned above. The data series are presented as follows:

- §3.2 Population (State level)
- §3.3 Urbanized Population
- §3.4 Population (Substate level)
- §3.5 Total Money Income, Per Capita Income (State level)
- §3.6 Per Capita Income (Substate level)
- §3.7 Personal Income
- §3.8 Net State and Local Taxes
- §3.9 State Individual Income Taxes
- §3.10 Federal Individual Income Tax Liabilities
- §3.11 Income Tax Amount (State level)
- §3.12 Adjusted Taxes (Substate level)
- §3.13 Intergovernmental Transfers (Substate level) .

Correlations between the different data series are discussed in §3.14 and §3.15.

A brief overview of the steps for developing error models will now be given. More technical points appear in §3.1-3.5 below.

Step 1 Extensive notation is necessitated by the large number of data series and error components. The logic of the notation is explined in §3.1.2 and §3.16 provides a glossary.

Step 2 The definition of each data series is considered and the data sources and procedures used to derive the data series are described.

Step 3 In constructing error models we rely heavily on evidence about the accuracy and relevance of the data sources and procedures. This evidence may come in various forms, including evaluations of data sources (such as reinterview studies), evaluation of procedures used (such as the assessment of methodology for estimating change in population since the 1970 census), and comparisons of the data series with alternative estimates of the parameters (such as use of demographic analysis to estimate population undercount in the 1970 census or use of special censuses to evaluate population updates); see Census (1976c), Census(1973c, no.21), and Siegel et al.(1977) for details. Errors in the estimates arise from numerous sources and different sets of evidence may relate to different components of error. For example, the error in the estimate of total population in 1976 arises from both error in the 1970 census counts and error in the estimates of population change from 1970 to 1976. To efficiently utilize the different pieces of evidence we obtain decompositions of the total error into the component pieces.

Step 4 Typically the evidence about data quality is incomplete, perhaps because some components of error have not been evaluated or because the evidence is not fine enough for our purposes. Different kinds of assumptions may be necessary to allow generalization from the evidence at hand. For example, in instances where the quality of the estimates has been studied only for a small set of governmental units, some assumptions are needed to extend inferences to the whole set of units. One kind of assumption often used is that of uniformity of error rates over homogeneous subgroups. For example, the net population undercoverage rates for different races in a state will be assumed uniform over all counties in the state.

Step 5 As a culmination of the four steps outlined above, explicit stochastic models for errors and their means, variances, and covariances are derived.

§3.1.1 Notation

"True values" (or parameter values) of the DE's are denoted by subscripted capital letters; estimates of the DE's are distinguished by a ^ , for example Z_i , $\hat{Z}_i$. Generally, subscripts i,j, and k refer to states, counties, and sub-county areas respectively. On occasion the data elements will be considered explicitly as functions of time t = 1,2,...10 measured on a scale of half-year periods beginning with 1/1/72 - 6/30/72. We will specify time by writing $Z_i(t)$ or $\hat{Z}_i(t)$. Times t = 1,2,3,5 correspond to the terminations of Entitlement Periods (EP's) 1 - 4, while times t = 7,9,10 correspond to the terminations of EP's 5 - 7. The uneven lengths of the EP's necessitates this awkwardness.

Lower case letters with appropriate subscripts denote deviations of estimated from true values, i.e.

$$z_i(t) = \hat{Z}_i(t) - Z_i(t) \quad .$$

The term "relative error" will be used for expressions of the form $z_i(t)/Z_i(t)$. As usual, a dot "." in place of a subscript denotes summation over the subscript.

Deviations of individual relative errors about group values are important for applying the delta method and it is convenient to have notation for means and variances of these deviations. The notation $\mu_Z(i;t)$, $\sigma_Z^2(i;t)$ is usually defined as follows:

$$\mu_Z(i;t) = E(z_i(t)/Z_i(t) - z.(t)/Z.(t))$$

(3.1)

$$\sigma_Z^2(i;t) = \mathrm{Var}(z_i(t)/Z_i(t) - z.(t)/Z.(t)) \quad .$$

Unfortunately the allocation formulas are so complicated that for some data elements Z the quantities μ_Z and σ_Z^2 must be defined by formulas other than (3.1). However, these definitions will usually follow the spirit, if not the fact, of (3.1). Precise definitions of σ_Z^2 and μ_Z are given for each data element Z in the appropriate subsections of this chapter.

Similar notation is used to handle covariances between estimates of data elements. For data elements $Y_i(t)$ and $Z_{i'}(t)$ the notation $\sigma_{YZ}(i,i';t)$ is usually defined by

(3.2) $$\sigma_{YZ}(i,i';t) = \mathrm{Cov}(y_i(t)/Y_i(t) - y.(t)/Y.(t), \; z_{i'}(t)/Z_{i'}(t) - z.(t)/Z.(t)) \quad .$$

As with (3.1), formula (3.2) is only suggestive, and precise definitions of σ_{YZ} will be given below in the respective subsections for notation.

The error in the estimate of change in a data element Z_i is denoted by $\Delta_Z(i;t) = z_i(t) - z_i(1) = \hat{Z}_i(t) - \hat{Z}_i(1) - (Z_i(t) - Z_i(1))$. Such an error will be referred to as "timeliness error".

A glossary (§3.16) is included to facilitate identification of the notation introduced in this chapter.

§3.1.2 Definition and Data Sources

The meaning of "true value" or parameter value of each data element will be as indicated in the GRS legislation*. For example,

---

* A transcript of the legislation appears in U.S. Statutes at Large, vol. 86.

state population P_i is "determined on the same basis as resident population is determined by the Bureau of the Census for general statistical purposes" (Public Law 92-512, Title I, sec.109(a)(1)).

Technology and society do not permit instantaneous population counts, and the time reference for $P_i(t)$ must be specified. The legislation requires only that the data used to estimate the parameter values "...shall be the most recently available data provided by the Bureau of the Census or the Department of Commerce, as the case may be" (sec.109(a)(7)(A)). For the purposes of this study it is assumed that the times referenced by the parameter values and the times referenced by the estimates coincide. This viewpoint tends to understate errors in estimates because no timeliness error will arise if no attempt is made to account for change over time.

For example, $\hat{P}_i(1)$ refers to April 1, 1970 population while $\hat{P}_i(3)$ refers to July 1, 1972 population. The mean square deviation of $\hat{P}_i(1)$ from its target value is less than the mean square deviation of $\hat{P}_i(3)$ from its target value because $\hat{P}_i(3)$ must account for additional sources of variation, namely births, deaths, and net migration from 4/1/70 to 6/1/72. The increase in mean square deviation for $\hat{P}_i$ arises solely from the attempt to account for change in population size over time. If the Bureau of Census had selected April 1, 1970 as the reference point for $\hat{P}_i(3)$ then the mean square deviation of $\hat{P}_i(3)$ would be no larger than that of $\hat{P}_i(1)$. From our perspective, timeliness error can only arise from inaccurate updating and not from failure to update the data to account for change over time. Thus, although the DE urbanized population $\hat{U}_i(t)$ was never revised to account for births, deaths, or migration and refers to April 1, 1970 urbanized population for each EP, we model this data series as having no timeliness error.

For each data element the procedure and data used to estimate the parameter value will be described generally with references to more complete descriptions provided. Listings of all the data elements for entitlement periods 1-9 are given in Office of

Revenue Sharing (1973; 1974a; 1974b; 1975; 1976; 1977; 1977b). On occasion references will be given for other listings of specific data elements. These references contain additional explanation of how the data was collected and analyzed.

§3.1.3 - 3.1.4 Data Quality: Evidence, Assumptions, and Conclusions

The relative error of a data element may be expressed as the sum of component errors which arise from sources such as non-response bias, response variance, sampling variance (see Cochran 1977, ch.3; Hansen, Hurwitz, and Bershad 1961). The error components are considered either as fixed effects or as random effects according to our conception of the nature of the components and our evidence concerning the component errors. For example, state population undercoverage rates have been extensively studied and are treated as fixed effects. If the census were repeated under identical conditions, we would expect state undercoverage rates to remain unchanged. On the other hand, errors in estimates of population change since the last census are treated primarily as random effects because they have been less extensively studied and because we have no evidence about directional biases. If the estimates of population change were redone under identical conditions and the collection of the data used for these estimates was also repeated then we would expect no systematic trends for the errors.

Consider the error of the estimate of total population on April 1, 1970, $p(1) = \hat{P}(1)-P(1)$. From the statistical viewpoint there are a variety of ways of viewing the unobservable quantity $p(1)$ Before the 1970 census was carried out we would think of $\hat{P}(1)$ as being random; in fact we might have known much about its distribution. The acceptability of the 1970 census program might have depended on known properties of the distribution of $p(1)$. For example, the program might have been unacceptable if the variance of $p(1)/P(1)$ was greater than 10^{-4} or if the expectation of $p(1)/P(1)$ was greater in absolute value than 10^{-2} . Such knowledge about the distribution would be gotten from, among other sources, retrospective evaluations of previous censuses and from pretests of the 1970 census.

After the 1970 census has been taken and the estimate $\hat{P}(1)$ formed the value of $p(1)$ is of course not known. Working in a non-Bayesian framework we think of $p(1)$ as fixed and consider its estimation. This estimation is based on new data, obtained by reenumerative studies (such as reinterviews), matching studies (such as Medicare check records), and analytical techniques (such as demographic analysis); see Census (1976c) and Siegel et al. (1977). Denote the estimate obtained from this additional data by $\tilde{p}(1)$. Knowledge of $\tilde{p}(1)$ permits retrospective estimation of $P(1)$, e.g. by $\tilde{P}(1) = \hat{P}(1) - \tilde{p}(1)$. Generally, the symbol $\tilde{}$ denotes a "second generation" estimate based on retrospective analysis of the original estimate (denoted by $\hat{}$).

We now consider the nature of the probability underlying (3.1) and (3.2). Suppose state i population $P_i(1)$ were to be estimated again solely on the basis of a hypothetical data collection and analysis stochastically independent but identical to that underlying $\hat{P}_i(1)$. Some moments of interest are

(3.3) $$\mu_P(i;1) = E\ p_i(1)/P_i(1) - p.(1)/P.(1)$$

and

(3.4) $$\sigma_P^2\ (i;1) = \mathrm{Var}(p_i(1)/P_i(1) - p.(1)/P.(1)) \quad .$$

To estimate these moments two different sources of information are used: (i) partial specification of the underlying stochastic model, and (ii) knowledge of the estimates $\tilde{p}_i(t)$. For example, (i) tells us that sampling variance is zero and response variance is negligible while (ii) provides estimates of undercoverage biases. It is from this "hypothetical" perspective that the estimates of the moments $\mu_P(i;1)$ and $\sigma_P^2(i;1)$ given in §3.2.5 below are to be interpreted. The specifications are based on evidence about (i) and (ii) given in §3.2.3 and assumptions about (i) and (ii) given in §3.2.4. Of course the values $\mu_P(i;1)$ and $\sigma_P^2(i;1)$ are

the same as they would be before the census was carried out.*
Thus the probability underlying (3.1) arises either in a "hypothetical" way or, equivalently, in a "retrospective" manner.

Another statistical perspective is the prospective rather than the retrospective or hypothetical. If the GRS program were going to be continued into the future then interest would focus on future values, say at time t' . Since the estimate $\hat{P}_i(t')$ has not been formed, $p_i(t')$ is random. If the estimator $\hat{P}_i(t')$ has a probability distribution similar to that of the estimator $\hat{P}_i(t)$, then knowledge of $\tilde{p}(t)$ together with partial specification of an underlying stochastic model provide for estimation of $\mu_p(i;t')$ and $\sigma_p^2(i;t')$.

The perspectives above have been considered from the non-Bayesian point of view. The Bayesian point of view is conceptually simpler but similar in practice. For the Bayesian and non-Bayesian alike, the differential error $p_i(1)/P_i(1) - p.(1)/P.(1)$ will never be precisely known. However the Bayesian will always represent his ignorance (or knowledge) about $p_i(1)/P_i(1) - p.(1)/P.(1)$ by a posterior probability distribution. At any point in time, the posterior distribution is conditional upon all available information. For example, before the 1970 census was carried out the posterior distribution would be conditional upon retrospective evaluations of previous censuses and pretests of the 1970 census. After the 1970 census was carried out the posterior distribution would be conditional upon the observed census counts as well as new data obtained by reenumerative studies, matching studies, and analytical methods. Thus the Bayesian and non-Bayesian perspectives are operationally similar. In particular, classification of error components as fixed or random components depends upon the judgement of the modeller, whether Bayesian or non-Bayesian.

---

* The estimates $\tilde{\mu}_p(i;1)$ and $\tilde{\sigma}_p^2(i;1)$ will reflect assumptions and assessment of the evidence at hand, and thus $\tilde{\mu}_p(i;1)$ and $\tilde{\sigma}_p^2(i;1)$ will not be the same.

Similarities between Bayesian and non-Bayesian perspectives notwithstanding, the perspective adopted is the non-Bayesian "retrospective" or "hypothetical" prospective view described in the third paragragh of this subsection.

Error models for sub-state estimates of population and per capita income will only be constructed for counties in New Jersey and for places and municipalities in Essex County, New Jersey. Although the procedures described to derive the error models are general, to derive explicit error models for all sub-state units would require a large amount of computing and is not done in this research.

§3.1.5 Relevant Distribution and Approximation

From the discussions on data quality (subsections 3,4) numerical values for the moments (3.1) and (3.2) (or explicit algorithms for determining them) are extracted. Several approximations will be required for the following quite different reasons:

(1.) Moments of functions of random variables are considered.

(2.) The evidence about data quality is inconclusive or incomplete.

(3.) Simplicity in the model is desired.

To get the flavor for construction of error models and assessment of the accuracy of data, the reader is encouraged to read at least section 3.2, on state-level population estimates. Other sections should be read as interest dictates. The results of the error models will be used in later chapters, but references will be made to appropriate parts of this chapter as their use arises.

§3.2 Population (State level)

§3.2.1 Notation

The population of a state is partitioned into the "white" population, denoted by subscript w, and the "blacks and other" (hereafter "blacks"), denoted by subscript b. Quantities of interest are

True (estimated) total population of U.S. at time t	$P(t)$ $(\hat{P}(t))$
True (estimated) population of state i at time t	$P_i(t)$ $(\hat{P}_i(t))$
True (estimated) white population of state i at time t	$P_{wi}(t)$ $(\hat{P}_{wi}(t))$
True (estimated) black population of state i at time t	$P_{bi}(t)$ $(\hat{P}_{bi}(t))$

By definition $P_i(t) = P_{wi}(t) + P_{bi}(t)$, $P(t) = P.(t)$ and by construction $\hat{P}_i(t) = \hat{P}_{wi}(t) + \hat{P}_{bi}(t)$, $\hat{P}(t) = \hat{P}.(t)$. Errors in GRS estimates of population size are denoted $p(t) = \hat{P}(t) - P(t)$; $p_i(t) = \hat{P}_i(t) - P_i(t)$; $p_{wi}(t) = \hat{P}_{wi}(t) - P_{wi}(t)$; $p_{bi}(t) = \hat{P}_{bi}(t) - P_{bi}(t)$ with estimates of errors denoted by $\tilde{p}(t)$; $\tilde{p}_i(t)$; $\tilde{p}_{wi}(t)$; $\tilde{p}_{bi}(t)$. The mean and variance of $p_i(t)/P_i(t) - p(t)/P(t)$ are denoted by $\mu_P(i;t)$ and $\sigma_P^2(i;t)$ respectively. The covariance of $p_i(t)/P_i(t) - p(t)/P(t)$ with $p_j(t)/P_j(t) - p(t)/P(t)$ is defined by $\sigma_{PP}(i,j;t)$.

Net undercoverage rates for the populations are defined by

$$A_i = E(-\, p_i(1)/P_i(1))$$

$$A_{wi} = E(-\, p_{wi}(1)/P_{wi}(1))$$

$$A_{bi} = E(-\, p_{bi}(1)/P_{bi}(1)) \quad \text{and}$$

$$A = E(-\, p(1)/P(1))$$

and their estimates are denoted by $\tilde{A}_i$, $\tilde{A}_{wi}$, $\tilde{A}_{bi}$, $\tilde{A}$. Errors in estimates of net increase are

$$\Delta_p(i;t) = p_i(t) - p_i(1) = \hat{P}_i(t) - \hat{P}_i(1) - (P_i(t) - P_i(1))$$
$$\Delta_p(t) = p(t) - p(1) = \hat{P}(t) - \hat{P}(1) - (P(t) - P(1)) \quad .$$

§3.2.2 Definition and Data Sources

By law, $\hat{P}_i$ is "determined on the same basis as resident population is determined by the Bureau of the Census for general statistical purposes" (Public Law 92-512, Title I, sec.109(a)(1)). For $t = 1,2$ $\hat{P}_i(t)$ is the April 1970 census count. For $t > 2$ the census counts were updated to account for births, deaths and net migration (for more detailed information see Census, 1974b No.520; Census 1976a No.640; SRI 1974a pp.III-6 ff). The estimates of state population used by GRS for EP 3-7 are published by the Census Bureau in Current Population Reports Series P-25 (see Census 1973b):

t	EP	Census Report	$P_i(t)$ is Population as of
1,2	1,2		April 1, 1970
3,5	3,4	No 488 Table, col. 1	July 1, 1972
7	5	No 508 Table, col. 1	July 1, 1973
9,10	6,7	No 615 Table 1, col. 2	July 1, 1973

Table 3.1

Target Variables - State Population

§3.2.3 Data Quality: Evidence

A major source of error in state population estimates is net undercount in the census. The Bureau of the Census estimated that nationally 7.7% of the blacks and 1.9% of the whites were not enumerated in the 1970 census (Siegel 1975b). Additional errors in state population estimates arise from errors in the reporting and classification of state of residence. Recent work by the Census (Siegel et al. 1977) provides a variety of estimates of net undercoverage rates A_i, A_{wi}, A_{bi} obtained by demographic analysis. This technique entails estimating the 1970 populations of states from data <u>other</u> than the 1970 census data. The difference between the census count and the demographic estimate is used to estimate the net undercount. Several strong assumptions were necessary to carry out this analysis, and varying the assumptions led to a whole set of alternative estimates of net undercoverage. In §3.2.4 below we select one set of estimates to estimate each of A_i, A_{wi}, and A_{bi}.

Another source of error is timeliness, that is, change in $P_i(t)$ as a function of t incorrectly accounted for by $\hat{P}_i(t)$. In EP's 3,4, and 5 the GRS estimates $\hat{P}_i$ were "provisional" Census estimates (say ${}_{prov}\hat{P}_i(t)$) and not the later "revised" estimates (say ${}_{rev}\hat{P}_i(t)$). The revisions reflect minor changes in the updating procedure as well as changes in the data input to the procedure. When $\hat{P}_i(t) = {}_{prov}\hat{P}_i(t)$ (i.e. EP's 3,4,5) one can obtain the following useful decomposition of net increase:

$$(3.5)\quad \Delta_p(i;t) = \underbrace{[{}_{prov}\hat{P}_i(t) - {}_{rev}\hat{P}_i(t)]}_{\text{revision error}} + \underbrace{[{}_{rev}\hat{P}_i(t) - P_i(t)) - (\hat{P}_i(1) - P_i}_{\text{updating error}}$$

No estimate of correlation between the revision error and the updating error was made. It is hypothesized that these error are uncorrelated*.

---

* This hypothesis could be tested by comparing provisional and revised population estimates ${}_{prov}\hat{P}_i(t)$, ${}_{rev}\hat{P}_i(t)$ against census counts, say $P_i'(t)$, for t corresponding to a year in which a census was taken. If $P_i'(t)-P_i(t)$ and $\hat{P}_i(1)-P_i(1)$ can be estimated from evidence about 1970 net state undercoverage rates, correlation could be estimated from the data on ${}_{prov}\hat{P}_i(t) - {}_{rev}\hat{P}_i(t)$ and ${}_{rev}\hat{P}_i(t) - P_i'(t)$.

The revision error is actually rather small, e.g. for the 1972 estimates the average percent absolute revision was 0.4% (Census, 1974b, No.520, p.14).

For EP's 3-7 updating error must be considered. The accuracy of the updating procedure is discussed in Census (1974b, No.520). After the 1970 census was taken, the method used in the 1960's to update the 1960 census was found to exhibit biases for certain regions of the country. The procedure was then modified for use in the 1970's. As a test of accuracy the modified procedure was applied to the 1960 data to estimate the 1970 population. The average per cent absolute deviation was 1.2% of the 1970 population size (Census 1974b, No.520, Table G).

§3.2.4 Data Quality: Assumptions and Conclusion

As mentioned in §3.1, the state undercoverage rates are considered as fixed effects. Census studies (Siegel et al., 1977) present estimates $\tilde{A}_i$, $\tilde{A}_{wi}$, $\tilde{A}_{bi}$ of the state undercoverage rates. The estimates used in the present research will be those designated respectively SOR-3-2-WCF-2-BACF-2, SOR-3-2-WCF-2, and SOR-3-2-WCF-2-BACF-2; (ibid. Appendix F, Tables F-1 col. 7, F-4 col. 3, and F-7 col. 6). These estimates were among the "preferred" estimates in Siegel et al. (1977), to which the reader is referred for more discussion.

Biases in the updating procedures used in the 1960's were detected by analyzing the 1970 census counts. The procedures were accordingly modified for use in the 1970's to reduce biases (Census 1974b, No.520, p.15). Consequently the relatively weak assumption is made

$$E(\Delta_P(i;t)/P_i(t) - \Delta_P(t)/P(t)) = 0 \quad . \tag{3.6}$$

The remainder of this section develops estimates of the variance of $\Delta_P(i;t)/P_i(t) - \Delta_P(t)/P(t)$. The variance of $\Delta_P(i;t)$ is the variance of the revision error plus the variance of the updating error (see (3.5)). Reason suggests that the variance of the updating error is an increasing function of time, but the specific form of the function is not known (ibid, p.15).

A plausible way to model the updating variance is to assume it is a linear function of time. For on one hand, methodological error should get worse over time as assumptions underlying the updating procedure become less valid. This might lead one to suspect that the variance was a convex function of time. On the other hand, improved data and analysis, such as special censuses and revisions of data used in previous censuses suggest that the increase in variance over time might not always be so rapid. Since the updating spans only three years at most, a linear function should give a reasonably good fit.

Both updating and revision errors are composed of a myriad of much smaller errors, so the central limit theorem motivates the assumption that they are normally distributed. Since $E(\Delta_P(i;t)/P_i(t) - \Delta_P(t)/P(t)) = 0$, with the assumption of normality, (2.25) implies that the mean square error is $\pi/2$ times the square of the mean absolute error. The mean absolute revision error, when present, is assumed to be 0.004 times the population size (see next to last paragraph in §3.2.3). Since the mean absolute updating error after 10 years is assumed to be 0.012 times the population size, the mean square updating error after m years is $(\pi/2)(0.012)^2(m/10)$. In contrast to the states, it is assumed that $Var(\Delta_P(t)/P(t))$ is negligible so we have:

$$\mathrm{Var}(\Delta_P(i;t)/P_i(t) - \Delta_P(t)/P(t)) = \begin{cases} 0 & t = 1,2 \\ (\pi/2)((0.004)^2 + 0.2(0.012)^2) & t = 3,5 \\ (\pi/2)((0.004)^2 + 0.3(0.012)^2) & t = 7 \\ (\pi/2)(0.4)(0.012)^2 & t = 9 \\ (\pi/2)((0.004)^2 + 0.5(0.012)^2) & t = 10 \ . \end{cases}$$

The assumption that $\mathrm{Var}(\Delta_P(t)/P(t))$ is negligible is justified because most of the error in updating and revision comes from error in estimating net internal migration. This error is negligible on a national basis because errors in interstate migration are irrelevant*.

The rationale for modelling the updating variances as constants across all states is the belief that procedural errors and biases dominate sampling variance as contributors to updating error. The Census studies (Census, 1974b, No.520, Table G) support this view: the average absolute relative updating error for the procedure used in the 1960's to predict 1970 populations was .0175 for "large" states (greater than 4.0 million inhabitants), .0259 for "medium-sized" states (population between 1.5 and 4.0 million) and .0118 for "small" states (population under 1.5 million).

To estimate $\sigma_{PP}(i,j;t) = \mathrm{Cov}(p_i(t)/P_i(t) - p(t)/P(t)\ ,\ p_j(t)/P_j(t) - p(t)/P(t))$ we will assume that $\sigma_{PP}(i,j;t) = \sigma_{PP}$ is constant for all $i \neq j$. This is reasonable because the variances σ_P^2 are constant for all states. Let $w_i = P_i(t)/P(t)$ and notice

$$\begin{aligned} &\Sigma\, w_k(p_k(t)/P_k(t) - p(t)/P(t)) \\ &= \Sigma \frac{P_k(t)}{P(t)} \left(\frac{p_k(t)}{P_k(t)} - \frac{p(t)}{P(t)}\right) \\ &= \Sigma \frac{p_k(t)}{P(t)} - \Sigma \frac{P_k(t)}{P(t)} \frac{p(t)}{P(t)} \\ &= \frac{p(t)}{P(t)} - \frac{p(t)}{P(t)} = 0 \ . \end{aligned}$$

---

* For this argument illegal immigrants are ignored.

It follows that

$$0 = \mathrm{Var}(\Sigma\, w_k(p_k(t)/P_k(t) - p(t)/P(t)))$$

$$= \sigma_P^2 \,\Sigma\, w_k^2 + \sigma_{PP} \sum_{k\neq\ell}\Sigma\, w_\ell w_k$$

hence (because $\Sigma\, w_k = 1$)

$$\sigma_{PP} = -\sigma_P^2 \,\Sigma\, w_k^2 \Big/ \sum_{k\neq\ell}\Sigma\, w_\ell w_k \tag{3.7}$$

$$= -\sigma_P^2 \,\frac{\Sigma\, w_k^2}{1-\Sigma\, w_k^2} \quad .$$

§3.2.5 Relevant Distributions and Approximations

Chapter 4 makes extensive use of $\mu_P(i;t)$ and $\sigma_P^2(i;t)$, the moments of $p_i(t)/P_i(t) - p(t)/P(t)$. To estimate these quantities the following lemma is helpful.

Lemma 3.0 The following identity holds:

$$p_i(t)/P_i(t) - p(t)/P(t) = \Delta_P(i;t)/P_i(t) - \Delta_P(t)/P(t) + p_i(1)/P_i(t) - p(1)/P(t) \quad . \tag{3.8}$$

Proof: Note that

$$p(t)/P(t) = [p(t) - p(1) + p(1)]/P(t)$$

$$= \Delta_P(t)/P(t) + p(1)/P(t)$$

and a similar relation holds for $p_i(t)/P_i(t)$. The conclusion is immediate.

*****

A reasonable estimate of $\mu_P(i;t)$ is

$$\tilde{A}\,\hat{P}(1)/\hat{P}(t) - \tilde{A}_i\,\hat{P}_i(1)/\hat{P}_i(t) \quad .$$

The estimate is obtained from (3.8) by replacing Δ_P by its expectation, which is zero. Further, $\tilde{A}$, $\tilde{A}_i$, $\hat{P}(1)/\hat{P}(t)$, $\hat{P}_i(1)/\hat{P}_i(t)$ are estimates of A , A_i , $P(1)/P(t)$, $P_i(1)/P_i(t)$ with small mean square error.

The values of $\tilde{A}$, $\tilde{A}_i$ are tabulated in Appendix A.

The quantity $\sigma_P^2(i;t)$ is estimated by the bound for

$$\mathrm{Var}(\,\Delta_P(i;t)/P_i(t) - \Delta_P(t)/P(t))$$

derived in §3.2.4.

From (3.7) the covariance $\sigma_{PP}(i,j;t)$ is estimated by

$$\tilde{\sigma}_{PP}(i,j;t) = \frac{-\tilde{\sigma}_P^2(j;t)\,\sum_k (\hat{P}_k(t)/\hat{P}.(t))^2}{1 - \sum_k (\hat{P}_k(t)/\hat{P}.(t))^2} \qquad i \neq j \quad .$$

Note that $\sigma_P^2(j;t)$ does not really depend on j , being constant for all states.

§3.3 Urbanized Population

The urbanized population of a state is partitioned into "white" population, denoted by subscript w , and the "blacks and others" (hereafter "blacks") denoted by subscript b . Let us define

True (estimated) urbanized population of state i at time t	$U_i(t)$ $(\hat{U}_i(t))$
True (estimated) white urbanized population of state i at time t	$U_{wi}(t)$ $(\hat{U}_{wi}(t))$
True (estimated) black urbanized population of state i at time t	$U_{bi}(t)$ $(\hat{U}_{bi}(t))$
True (estimated) total urbanized population of U.S. at time t	$U(t)$ $(\hat{U}(t)$.

By definition $U_i(t) = U_{wi}(t) + U_{bi}(t)$, $U(t) = \Sigma U_i(t)$ and by construction $\hat{U}_i(t) = \hat{U}_{wi}(t) + \hat{U}_{bi}(t)$, $\hat{U}(t) = \Sigma \hat{U}_i(t)$. Errors in GRS estimates of urbanized population are denoted by

$u_i(t) = \hat{U}_i(t) - U_i(t)$; $u_{wi}(t) = \hat{U}_{wi}(t) - U_{wi}(t)$;

$u_{bi}(t) = \hat{U}_{bi}(t) - U_{bi}(t)$; $u(t) = \hat{U}(t) - U(t)$. Estimates of these errors are denoted by $\tilde{u}_i(t)$; $\tilde{u}_{wi}(t)$; $\tilde{u}_{bi}(t)$; $\tilde{u}(t)$. The mean and variance of $u_i(t)/U_i(t) - u(t)/U(t)$ are denoted by $\mu_U(i;t)$ and $\sigma_U^2(i;t)$ respectively. The covariance of $u_i(t)/U_i(t) - u(t)/U(t)$ with $u_j(t)/U_j(t) - u(t)/U(t)$ is denoted by $\sigma_{UU}(i,j;t)$.

For $U_i(t) > 0$, undercoverage rates for the urbanized populations are denoted by

$$A_{Ui}(t) = E(-u_i(t)/U_i(t))$$

$$A_U(t) = E(-u(t)/U(t))$$

with estimates $\tilde{A}_{Ui}(t)$ and $\tilde{A}_U(t)$.

§3.3.2 Definition and Data Source

According to legislation "urbanized population means the population of any area consisting of a central city or cities of 50,000 or more inhabitants (and of the surrounding closely settled territory for such city or cities) which is treated as an urbanized area by the Bureau of the Census for general statistical purposes" (Public Law 92-512, Title I, sec.109 (a)(2)). In 1973 the Census Bureau changed its definition of urbanized area and the definition of urbanized population changed accordingly. The new definition of urbanized area is similar to the old but more complicated. It is described in Office of Revenue Sharing (1976,pp.xi,xii).

The estimates of urbanized population are based on complete enumeration in the 1970 Census of Population and subsequent classification of population density. For more discussion see SRI (1974a, pp.38,39). The estimates were not updated, other than to account for the definitional change of urbanized area. Thus for all entitlement periods $U_i(t)$ refers to April 1, 1970 population.

§3.3.3 Data Quality: Evidence

Estimates of urbanized population are subject to errors arising from underenumeration and misclassification by the Census. According to the unpublished 1970 Census-CPS Match Study cited in Siegel (1975, p.9)

> Coverage rates for cities of 1,000,000 or more and cities of 50,000 to 1,000,000 were not significantly different from the national average...This study does not provide support to the commonly held view that coverage is poorer in large cities. It is reasonable to surmise that cities having heavy concentrations of blacks have higher undercoverage rates than areas with much smaller concentrations of blacks.

Misclassification error is small. From Census (1972a, Table 1, col.g) the probability of misclassification of a housing unit into the wrong minor civil division, census county division, or place is estimated to be smaller than 0.008.

§3.3.4 Data Quality: Assumptions and Conclusions

The undercoverage rates for urbanized state populations are considered as fixed effects. Extrapolation of substate undercoverage rates from state rates is hard. Siegel's remarks (see §3.3.3) suggest that it is not unreasonable to assume the undercoverage rates for white and black urbanized state populations are the same as those for the state populations. That is, for $U_{w,i}(t) > 0$, $U_{b,i}(t) > 0$ we assume

$$E(-u_{wi}(t)/U_{wi}(t)) = A_{wi} \tag{3.9}$$

$$E(-u_{bi}(t)/U_{bi}(t)) = A_{bi} \quad . \tag{3.10}$$

<u>Lemma 3.1</u> Under assumptions (3.9), (3.10)

$$A_{Ui}(t) = 1 - \frac{E(\hat{U}_{wi}(t) + \hat{U}_{bi}(t))}{E(\hat{U}_{wi}(t)/(1-A_{wi}) + \hat{U}_{bi}(t)/(1-A_{bi})} \quad . \tag{3.11}$$

<u>Proof</u>: First notice that from (3.9)

$$E\left(\frac{U_{wi}(t) - \hat{U}_{wi}(t)}{U_{wi}(t)}\right) = A_{wi}$$

so

$$E\,\frac{-\hat{U}_{wi}(t)}{U_{wi}(t)} = (A_{wi} - 1)$$

hence

$$U_{wi}(t) = E\,\hat{U}_{wi}(t)/(1-A_{wi}) \quad .$$

Similarly,

$$U_{bi}(t) = E\,\hat{U}_{bi}(t)/(1-A_{bi}) \quad .$$

By definition,

$$A_{Ui}(t) = 1 - \frac{E(\hat{U}_{wi}(t) + \hat{U}_{bi}(t))}{U_{wi}(t) + U_{bi}(t)} \quad .$$

Substituting for $U_{wi}(t)$, $U_{bi}(t)$ gives (3.11).

*****

Although undercoverage errors occur only during the census, it is neçessary to have $A_{Ui}(t)$ depend on t because the definition of urbanized area changed with t.

From (3.11) we estimate $A_{Ui}(t)$ by

$$\tilde{A}_{Ui}(t) = 1 - (\hat{U}_{wi}(t) + \hat{U}_{bi}(t))/[\hat{U}_{wi}(t)/(1-\tilde{A}_{wi}) + \hat{U}_{bi}(t)/(1-\tilde{A}_{bi})]$$

where $\tilde{A}_{wi}$, $\tilde{A}_{bi}$ are described in §3.2.4*. To estimate $A_U(t)$ we will use

$$\tilde{A}_U(t) = \Sigma \tilde{A}_{Ui}(t) \tilde{U}_i(t)/\Sigma \tilde{U}_j(t)$$

where $$\tilde{U}_i(t) = \hat{U}_i(t)/(1-\tilde{A}_{Ui}(t)) \quad .$$

It will now be argued that $Var(u_i(t)/U_i(t))$ is small. First, recall that the main source of random error in $\hat{U}_i$ arises from the small probability of misclassifying a household. The probability of misclassifying a household as urbanized or not urbanized is probably even less than the probability of misclassification into the wrong minor civil division, census county division, or place (.008) because urbanized areas are larger. Except for a few states the number of individuals not in urbanized areas is less than $5U_i$. Thus the number of people living outside an urbanized area subject to misclassification into an urbanized areas is $\leq 5U_i$, while the number of people living in an urbanized area subject to misclassification out of the urbanized area is $\leq U_i$. Notice that the variance of the net number of misclassifications is less than or equal to the

---

* Estimates $\tilde{A}_{bi}$ were not provided by Siegel et al. (1977) for Maine, New Hampshire, Vermont, or Wyoming because in each of these states the enumerated black population numbered less than 10,000. In deriving estimates $\tilde{A}_{Ui}$, regional undercoverage rates will be substituted for the missing $\tilde{A}_{bi}$.

sum of the variances of the number of either kind of misclassification. Assuming that misclassifications occur independently and that the average household size is 4 gives

$$\mathrm{Var}\ (\hat{U}_i - U_i) \le (5U_i(.008)(.992) + U_i(.008)(.992))4 < .20\ U_i\ .$$

Since $U_i > 50{,}000$ it follows that

$$\mathrm{Var}\ (u_i/U_i) < .2/U_i < 4\times 10^{-6}\ .$$

§3.3.5 Relevant Distributions and Approximations

To estimate $\mu_U(i;t)$ use $\tilde{A}_U(t) - \tilde{A}_{Ui}(t)$, where $\tilde{A}_U(t)$, $\tilde{A}_{Ui}(t)$ are described in §3.3.4. Values of $\tilde{\mu}_U(i;t)$ are tabulated in Appendix A. We have from §3.3.4 that $\sigma_U^2(i;t)$ is negligible, i.e. $\sigma_U^2(i;t) < 10^{-3}$. Thus $\sigma_{UU}(i,j;t) < 10^{-3}$ as well.

§3.4 Population (Substate level)

§3.4.1 Notation

Define

True (estimated) population of county j in state i at time t $\quad P_{ij}(t)\ (\hat{P}_{ij}(t))$

True (estimated) population of local area k in county j in state i at time t $\quad P_{ijk}(t)\ (\hat{P}_{ijk}(t))$

Errors in GRS estimates of substate populations are denoted by $p_{ij}(t) = \hat{P}_{ij}(t) - P_{ij}(t)$ and $p_{ijk}(t) = \hat{P}_{ijk}(t) - P_{ijk}(t)$ with estimates $\tilde{p}_{ij}(t)$ and $\tilde{p}_{ijk}(t)$. The means and variances of $p_{ij}(t)/P_{ij}(t) - p_{i.}(t)/P_{i.}(t)$ and $p_{ijk}(t)/P_{ijk}(t) - p_{i..}(t)/P_{i..}(t)$

are denoted by $\mu_P(i,j;t)$, $\sigma_P^2(i,j;t)$, $\mu_P(i,j,k;t)$, $\sigma_P^2(i,j,k;t)$. Net undercoverage rates are

$$A_{ij} = E -p_{ij}(1)/P_{ij}(1) \quad , \quad A_{ijk} = E -p_{ijk}(1)/P_{ijk}(1) ,$$

and their estimates are denoted by $\tilde{A}_{ij}$, $\tilde{A}_{ijk}$.

Errors in estimates of net increases are

$$\Delta_P(i,j;t) = p_{ij}(t) - p_{ij}(1) \quad , \quad \Delta_P(i,j,k;t) = p_{ijk}(t) - p_{ijk}(1) .$$

Define

$$\Delta_P(i,.;t) = \sum_j \Delta_P(i,j;t) \quad , \quad \Delta_P(i,.,.;t) = \sum_j \sum_k \Delta_P(i,j,k;t) .$$

It will be seen that because of different time references for $P_{ij}(t)$ and $P_i(t)$, $\Delta_P(i,.;t)$ does not equal $\Delta_P(i;t)$. However $\Delta_P(i,.;t)$ and $\Delta_P(i,.,.;t)$ are equal.

§3.4.2 Definition and Data Sources

Like state population $\hat{P}_i$, both $\hat{P}_{ij}$ and $\hat{P}_{ijk}$ are estimates of population "determined on the same basis as resident population is determined by the Bureau of the Census for general statistical purposes" (Public Law 92-512, Title I, sec. 109(a)(1)). For entitlement periods 1-5 both P_{ij} and P_{ijk} are based on the Census counts of April 1, 1970 resident population. In EP's 1-2,3-4 and 5 the estimates were revised to account for boundary changes as of December 31, 1971, 1972, and 1973 respectively; see Census (1975a) and SRI (1974a, appendix B-1).* In EP's 6-7 the estimates of population

---

* However the Census Bureau did not in general attempt to adjust for boundary changes the population estimates for places with 1970 population under 2500. For more discussion see SRI (1974a, appendix B-1).

were updated to refer to July 1, 1973 population with political boundaries of December 31, 1973. The estimates for EP's 6-7 are published by the Bureau of the Census in Census (1975b, nos.546-595) for all states except the District of Columbia. County (subcounty) updates were adjusted to be consistent with state (county) updates.

The following specification is adopted:

t	EP	$P_{ij}(t)$, $P_{ijk}(t)$ is population as of
1,2	1,2	April 1, 1970
3,5	3,4	April 1, 1970
7	5	April 1, 1970
9,10	6,7	July 1, 1973

Table 3.2

Target Variables - Substate Population

§3.4.3 Data Quality: Evidence

The sources of error considered for substate population estimates are census undercoverage and timeliness of the estimates. Much less is known about the undercoverage rates for substate populations than for state populations. In §3.4.4 a synthetic approach is taken to estimate sub-state undercoverage rates.

In EP's 4,5 no updating of estimates was performed aside from revisions to account for boundary changes as of December 31, 1972 and December 31, 1973. The author could not locate any published evidence about the effects of errors in the surveys of boundary changes upon estimates of substate population sizes.

The methods used to update the 1970 county population counts for EP 6 are similar to those used for states (see §3.2.2 above) The best updating procedures for counties in New Jersey in the 1960's had a mean absolute deviation of 3.5% after 10 years (Census 1973c no.21, Table 2, col.4). Special censuses taken in the 1970's suggest that the mean absolute deviation for current procedures is roughly one half that of those used in the 1960's (personal communication 1/16/78 with Fred Cavanaugh, Census Bureau).

For large subcounty units the updating error variance will be comparable to that of counties. Small areas are much worse. Census studies (ibid., Table D, col.1) of accuracy of the 1960s' updates indicates a mean absolute deviation of about 7% over ten years for updates of counties with less than 5,000 population, obtained by a predecessor of the updating procedure used in the 1970's in New Jersey. No data about the accuracy of the current substate population updates have been published.*

§3.4.4 Data Quality: Assumptions and Conclusions

A "synthetic" approach is taken to estimate A_{ij} , A_{ijk} ; see Shryock, Siegel et al. (1973, pp.326-327) and Gonzalez (1974, pp.46-50) for discussion of the synthetic method. The synthetic estimates will not necessarily be accurate for individual substate units but the overall picture should be indicative. Estimates $\tilde{A}_{ij}$, $\tilde{A}_{ijk}$ are obtained in the same manner as $\tilde{A}_{Ui}$ were obtained in §3.3.4. That is, the net undercoverage rates by race are assumed to be uniform over all county and subcounty units within a state. Thus we set

$$\tilde{A}_{ij} = 1 - (\hat{P}_{wij}(1) + \hat{P}_{bij}(1))/[\hat{P}_{wij}(1)/(1-\tilde{A}_{wi}) + \hat{P}_{bij}(1)/(1-\tilde{A}_{bi})]$$

and

$$\tilde{A}_{ijk} = 1 - (\hat{P}_{wijk}(1) + \hat{P}_{bijk}(1))/[\hat{P}_{wijk}(1)/(1-\tilde{A}_{wi}) + \hat{P}_{bijk}(1)/(1-\tilde{A}_{bi})] \quad .$$

* An extensive assessment of the accuracy of these updates has recently been completed by the National Research Council (1980).

Now consider timeliness. As a consequence of the specification of $P_{ij}(t)$, $P_{ijk}(t)$ (see Table 3.2)

$$\left.\begin{aligned} \Delta_P(i,j;t) &= 0 \\ \Delta_P(i,j,k;t) &= 0 \end{aligned}\right\} \qquad t = 1,2,\ldots,7 \quad .$$

As with state-level population updates, the assumption is made that biases in the substate population updates are uniform throughout the state:

$$(3.12) \qquad E\left(\frac{\Delta_P(i,j;t)}{P_{ij}(t)} - \frac{\Delta_P(i,\cdot;t)}{P_{i\cdot}(t)}\right) = 0 \qquad t = 9,10$$

$$(3.13) \qquad E\left(\frac{\Delta_P(i,j,k;t)}{P_{ijk}(t)} - \frac{\Delta_P(i,\cdot,\cdot;t)}{P_{i\cdot\cdot}(t)}\right) = 0 \qquad t = 9,10 \quad .$$

Estimates of the variance of the updates can be derived from the evidence of the preceding section using the method described in §3.2.4. Since the population updates for $t = 9,10$ estimate three years of population increase since the 1970 census, and the estimated mean relative absolute deviation for a ten-year county-level update is (.035)/2 , we will estimate the variance of $\Delta_P(i,j;t)/P_{ij}(t) - \Delta_P(i,\cdot;t)/P_{i\cdot}(t)$ by $(\pi/2)(.3)(.0175)^2$ for $t = 9,10$.

Similarly, for local areas in Essex County, New Jersey the variance of $\Delta_P(i,k;t)/P_{ijk}(t) - \Delta_P(i,\cdot,\cdot;t)/P_{i\cdot\cdot}(t)$ is estimated for $t = 9,10$ by

$$5.8 \times 10^{-4} \times (\pi/2)(.3)(.035)^2 \qquad \text{if } P_{ijk}(t) \leq 5000$$

$$1.4 \times 10^{-4} \times (\pi/2)(.3)(.0175)^2 \qquad \text{if } P_{ijk}(t) > 5000 \quad .$$

The remainder of this subsection, §3.4.4, is devoted to showing that certain covariances are negligible. The conclusions are stated below as (3.16) and (3.17). In the following, the index "N" is large. All of the random variables depend implicitly on N but the constants do not. The reader is referred to §3 of chapter 2 for an earlier discussion of the interpretation of N. In keeping with the orders of approximation arising from the delta method (theorem 2.1), a first moment will be neglected if $o(N^{-1/2})$ and a second moment will be neglected if $o(N^{-1})$. Some discussion of the size of these neglected terms appears in §4 of chapter 4.

In the next lemma each random variable ε (possibly with subscripts) has the following properties:

(i) $$\varepsilon = O_p(N^{-1/2})$$

(ii) $$E\varepsilon = o(N^{-1/2})$$

(iii) $$\mathrm{Var}(\varepsilon) = O(N^{-1})$$

(iv) $$E(\varepsilon - E\varepsilon)^4 = o(N^{-1})$$

<u>Lemma 3.2</u>

Let $\hat{X}$ and $\hat{X}_i$ be random variables, let Y and Y_i be constants such that

$$\hat{X} = Y + \varepsilon \quad ,$$
$$\hat{X}_i = Y_i + \varepsilon \quad ,$$

and

$$Y = \Sigma Y_i$$

and let ε and ε_i satisfy (i)-(iv) above. Further assume that ε and ε_i are independent and that $\hat{X}.$ is positive and bounded away from zero, and define

(3.15) $$\hat{Y}_i = \hat{X}_i \hat{X} / \hat{X}. \quad .$$

It follows that

$$\mathrm{Cov}\left(\frac{\hat{Y}_i}{Y_i} - \frac{\hat{X}}{Y}, \frac{\hat{X}}{Y}\right) = 0(N^{-3/2}) \quad .$$

Proof: Define the quantities A and B by

$$A = \frac{\hat{X}}{Y} = \frac{Y+\varepsilon}{Y}$$

and

$$B = \frac{Y}{Y_i} \frac{\hat{X}_i}{\hat{X}.} - 1 = \frac{Y}{Y_i} \frac{(Y_i+\varepsilon_i)}{(Y+\varepsilon.)} - 1 \quad .$$

Notice that

$$AB = \frac{\hat{Y}_i}{Y_i} - \frac{\hat{X}}{Y}$$

so that

$$\mathrm{Cov}\left(\frac{\hat{Y}}{Y_i} - \frac{\hat{X}}{Y}, \frac{\hat{X}}{Y}\right) = \mathrm{Cov}(AB,A).$$

Since ε and ε_i are independent, so are A and B, hence the equality $\mathrm{Cov}(AB,A) = E(B)\cdot \mathrm{Var}(A)$ holds. By the delta method $E(B) = 0(N^{-1/2})$ and by hypothesis $\mathrm{Var}(A) = 0(N^{-1})$. The conclusion is immediate.

*****

The form of $\hat{Y}_i$ in (3.15) is motivated by the adjustment of county updates to be consistent with state updates. To use the lemma replace $\hat{X}$ by $\hat{P}_i(9)$ and $\hat{Y}_i$ by $\hat{P}_{ij}(9)$. The assumption of independence between ε and ε_i is justified because the error in the unadjusted county updates arises principally from errors in estimating intrastate migration and errors in attributing births and deaths to the wrong county in the state, while the error in the state updates comes mainly from errors in estimating interstate migration. The lemma therefore implies

$$\mathrm{Cov}\left[\frac{\hat{P}_{ij}(t)}{P_{ij}(t)} - \frac{\hat{P}_{i}.(t)}{P_{i}.(t)} , \frac{\hat{P}_{i}.(t)}{P_{i}.(t)}\right] = O(N^{-3/2})$$

for $t = 9$. The case $t = 10$ is identical, and for $t < 9$ the substate errors are fixed effects, and hence the conclusion is true. In a similar manner we can show

$$\mathrm{Cov}\left[\frac{\hat{P}_{ij}(t)}{P_{ij}(t)} - \frac{\hat{P}_{i}.(t)}{P_{i}.(t)} , \frac{\hat{P}..(t)}{P..(t)}\right] = O(N^{-3/2}) .$$

Taken together, these equations yield

$$\text{(3.16)} \quad \mathrm{Cov}\left[\frac{\hat{P}_{ij}(t)}{P_{ij}(t)} - \frac{\hat{P}_{i}.(t)}{P_{i}.(t)} , \frac{\hat{P}_{i}.(t)}{P_{i}.(t)} - \frac{\hat{P}..(t)}{P..(t)}\right] = O(N^{-3/2}) .$$

Similar arguments lead to

$$\text{(3.17)} \quad \mathrm{Cov}\left[\frac{\hat{P}_{ijk}(t)}{P_{ijk}(t)} - \frac{\hat{P}_{i}..(t)}{P_{i}..(t)} , \frac{\hat{P}_{i}..(t)}{P_{i}..(t)} - \frac{\hat{P}...(t)}{P...(t)}\right] = O(N^{-3/2}) .$$

§3.4.5 Relevant Distribution and Approximations (for New Jersey)

In analogy with state populations (see §3.2.5) a reasonable estimate of $\mu_P(i,j;t)$ is

$$\tilde{\mu}_P(i,j;t) = \tilde{A}_i - \tilde{A}_{ij} \qquad t < 9$$

$$= \tilde{A}_i \frac{\hat{P}_{i}.(1)}{\hat{P}_{i}.(t)} - \tilde{A}_{ij} \frac{\hat{P}_{ij}(1)}{\hat{P}_{ij}(t)} \qquad t = 9,10$$

and a reasonable estimate of $\mu_P(i,j,k;t)$ is

$$\begin{cases} \tilde{A}_i - \tilde{A}_{ijk} & t < 9 \\ \tilde{A}_i \dfrac{\hat{P}_{i\cdot\cdot}(1)}{\hat{P}_{i\cdot\cdot}(t)} - \dfrac{\tilde{A}_{ijk}\hat{P}_{ijk}(1)}{\hat{P}_{ijk}(t)} & t = 9,10 \end{cases} .$$

For $t < 9$ both $\sigma_P^2(i,j;t)$ and $\sigma_P^2(i,j,k;t)$ are zero while for $t = 9,10$ $\sigma_P^2(i,j;t)$ and $\sigma_P^2(i,j,k;t)$ are estimated by the estimates given in §3.4.3. That is, $\sigma_P^2(i,j;t)$ is estimated by

$$(\pi/2)(.3)(.0175)^2 \qquad t = 9,10$$

and $\sigma_P^2(i,j,k;t)$ is estimated by

$$\begin{cases} (\pi/2)(0.3)(.0175)^2 & P_{ijk}(9) > 5000 \\ (\pi/2)(0.3)(.035)^2 & P_{ijk}(9) \leq 5000 \end{cases} .$$

Values of $\tilde{A}_i$, $\tilde{A}_{ij}$, $\tilde{A}_{ijk}$ are tabulated in Appendix A.

§3.5 Total Money Income, Per Capita Income (State level)

§3.5.1 Notation

Denote

True (estimated total money income of state i at time t	$M_i(t)$ $(\hat{M}_i(t))$
True (estimated) total national money income at time t	$M(t)$ $(\hat{M}(t))$
True (estimated) per capita income of state i at time t	$C_i(t)$ $(\hat{C}_i(t))$
True (estimated) national per capita income at time t	$C(t)$ $(\hat{C}(t))$.

Errors in estimates of total money income and per capita income are denoted by $m_i(t) = \hat{M}_i(t)-M_i(t)$; $m(t) = \hat{M}(t)-M(t)$; $c_i(t) = \hat{C}_i(t)-C_i(t)$; $c(t) = \hat{C}(t)-C(t)$. The mean and variance of $c_i(t)/C_i(t) - c(t)/C(t)$ are denoted respectively by $\mu_C(i;t)$ and $\sigma_C^2(i;t)$. The covariance of $c_i(t)/C_i(t) - c(t)/C(t)$ with $c_j(t)/C_j(t) - c(t)/C(t)$ is denoted by $\sigma_{CC}(i,j;t)$. Errors in estimates of net increase in per capita income and money are denoted by

$$\Delta_C(i;t) = c_i(t) - c_i(1)$$

$$\Delta_C(t) = c(t) - c(1)$$

$$\Delta_M(i;t) = m_i(t) - m_i(1)$$

$$\Delta_M(t) = m(t) - m(1) \quad .$$

§3.5.2 Definitions and Data Sources

Total money income is the sum of wage and salary income, net non-farm income, self-employment income, social security or railroad retirement income, public assistance income, and all other income such as interest, dividends, veteran's payments, pensions, alimony, etc. (SRI, 1974a, pp.III-20 ff.) The per capita income of an area is the mean amount of total money income received in a year by a person residing in the area.

The following specification for $M_i(t)$, $C_i(t)$ holds

t	EP	$M_i(t)$, $C_i(t)$ refer to calendar year
1,2	1,2	1969
3,5	3,4	1969
7	5	1969
9,10	6,7	1972

Table 3.3

Target Variables - State Money Income, State Per Capita Income

The basic data for estimating total money income was obtained from a 20% sample of the 1970 Census of Population. For EP's 1-5 total money income and per capita income estimates were computed from the population data and calendar year 1969 total money income data from the 1970 census. In EP's 6,7 the per capita income estimates were updated by a complicated procedure (Census 1975a No.575, p.III; Herriot 1978). For example, the EP 6 updates were based upon the following data sources: 1970 census, 1969 and 1972 federal income tax returns, and a special set of state money income estimates prepared by the Bureau of Economic Analysis. These data were used to estimate the change in money income from 1969 to 1972, which were then added to the estimate for 1969 to provide the EP 6 estimates of money income. These updated estimates of money income were then divided by updated population estimates to yield estimates of 1972 per capita income.

Because $P_i(t)$ and $M_i(t)$ do not necessarily refer to the same years, it is not in general true that $C_i(t) = M_i(t)/P_i(t)$. However the following relations do hold

$$(3.18) \qquad C_i(1) = M_i(1)/P_i(1) \qquad \hat{C}_i(1) = \hat{M}_i(1)/\hat{P}_i(1)$$

$$(3.19) \qquad C_i(9) = M_i(9)/P_i(9) \qquad \hat{C}_i(9) = \hat{M}_i(9)/\hat{P}_i(9) \ .$$

Since $C_i(t) = C_i(1)$ for EP's 1-5 and $C_i(t) = C_i(9)$ for EP's 6-7, expressions (3.18) and (3.19) are sufficient for decomposing $C_i(t)$ for all EP's.

§3.5.3 Data Quality: Evidence

Major sources of error include response bias, undercoverage bias, and updating errors. As a consequence of large sample sizes at the state level, sampling variance and response variance are negligible. Population undercoverage introduces a bias because the unenumerated populations have lower than average per capita income.

Evaluations of the accuracy of the updating procedures have not been reported. The accuracy could be investigated in the same way that the accuracy of the population updating procedures was studied (see §3.2.3 above). It is expected that such an investigation will be undertaken when the 1980 census is conducted.

The accuracy of income reporting in the 1970 census varies by source of income. Using 1969 Bureau of Economic Analysis data to derive benchmark estimates, M. Ono (1972) estimated that on a national basis virtually all wage or salary and non-farm income was reported, while only 82% of social security and 69% of public assistance income was reported. Ono also gives estimates of proportion of income reported by each state for each type of income. The accuracy of these estimates is not reported.

§3.5.4 Data Quality: Assumptions and Conclusions

It is useful to examine errors in money and population separately. For $t = 1,9$ formulas (3.18) and (3.19) give

$$\frac{c(t)}{C(t)} = [\frac{\hat{M}(t)}{\hat{P}(t)} - \frac{M(t)}{P(t)}]/(M(t)/P(t))$$

with similar expression for $c_i(t)/C_i(t)$. It is a direct application of the delta method (theorem 2.1) that

$\frac{c_i(t)}{C_i(t)} - \frac{c(t)}{C(t)}$ equals

$$\frac{m_i(t)}{M_i(t)} - \frac{m(t)}{M(t)} - (\frac{p_i(t)}{P_i(t)} - \frac{p(t)}{P(t)}) \qquad t = 1,9 \tag{3.20}$$

$$+ \ o_p(N^{-1/2}) \ .$$

Thus for $t = 1,9$ an estimate of $\mu_C(i;t)$ can be derived by estimating the mean of the leading terms of (3.20).

Recall that in §3.2.5 estimates were presented for $\mu_P(i;t) = E(p_i(t)/P_i(t) - p(t)/P(t))$. Now consider the mean of

$$\frac{m_i(t)}{M_i(t)} - \frac{M(t)}{M(t)}$$

which is identically equal to

$$\frac{m_i(1) + \Delta_M(i;t)}{M_i(t)} - \frac{M(1) + \Delta_M(t)}{M(t)} \quad .$$

It will be assumed that the relative biases in the updates of money income are uniform over the states, that is

$$(3.21) \qquad E\left(\frac{\Delta_M(i;t)}{M_i(t)} - \frac{\Delta_M(t)}{M(t)}\right) = 0 \ .$$

This is not an assumption about lack of biases, but rather that the differences in the relative biases are small enough to ignore. The assumption is subject to test when the 1980 census results are analyzed.

It follows that

$$(3.22) \qquad E\left(\frac{m_i(t)}{M_i(t)} - \frac{m.(t)}{M.(t)}\right) = E\left[\frac{m_i(1)}{M_i(1)}\,\frac{M_i(1)}{M_i(t)} - \frac{m.(1)}{M.(1)}\,\frac{M.(1)}{M.(t)}\right]$$

To evaluate the right hand side of (3.22) estimate $M_i(1)/M_i(t)$ and $M.(1)/M.(t)$ by $\hat{M}_i(1)/\hat{M}_i(t)$ and $\hat{M}.(1)/\hat{M}.(t)$. Estimates of the means of $m_i(1)/M_i(1)$ and $m.(1)/M.(1)$ are gotten easily from data in Ono (1972, Table 5, col.1). From the resulting estimates of the right hand side of (3.22) the estimates of $\mu_P(i;t)$ can be subtracted, providing an estimate $\tilde{\mu}_C(i;t)$ (tabulted in Appendix A).

Now consider $\sigma_C^2(i;t)$. Since sampling variance and response variance are negligible, $\sigma_C^2(i;t) = 0$ for EP's 1-5. For EP's 6-7 we must consider the variance of the updating procedure. Although the accuracy of the updating procedure for per capita income estimates has not been extensively investigated, crude upper bounds for the variance of the updating procedure can be derived. Since considerable effort was expended to derive the updates, it is reasonable to assume that the expected squared error introduced by updating is no greater than the squared error introduced if no updating had been done. In particular, we assume

$$(3.23)\quad E[\Delta_C(i;9)/C_i(9) - \Delta_C(9)/C(9)]^2 \le [(C_i(9)-C_i(1))/C_i(9) - (C(9)-C(1))/C(9)]^2 \quad .$$

Assuming that the left hand side of (3.23) is the same for all states we estimate the upper bound by averaging

$$\left[\frac{\hat{C}_i(9) - \hat{C}_i(1)}{\hat{C}_i(9)} - \frac{\hat{C}(9) - \hat{C}(1)}{\hat{C}(9)}\right]^2$$

over all states. This average is 0.0008. In light of (3.6) and (3.21)

$$E(\Delta_C(i;9)/C_i(9) - \Delta_C(9)/C(9)) = 0$$

and so we estimate that

$$Var(\Delta_C(i;9)/C_i(9) - \Delta_C(9)/C(9) \le 0.0008 \quad .$$

This upper bound may be close. The corresponding estimated variance of the population updating procedure for a three year update is 0.0009 (see §3.2.5) and we do not expect the per capita income updates to be substantially more accurate than the population updates. On the other hand, the fact that the estimated variance for the per capita income updates is less than that for population may indicate that our "upper bound" of .0008 is much too small.

§3.5.5 Relevant Distributions and Approximations

For EP's 1-5

$$\tilde{\sigma}_C^2(i;t) = 0 \qquad\qquad t = 1,2,3,5,7$$

while for EP's 6-7 we use the upper bound of .0008 to estimate

$$\tilde{\sigma}_C^2(i;t) = .0008 \qquad\qquad t = 9,10 \quad .$$

As discussed in §3.5.4 we will estimate $\mu_C(i;t)$ by

$$\frac{\hat{M}_i(1)}{\hat{M}_i(t)} \cdot E\left(\frac{m_i(1)}{M_i(1)}\right) - \frac{\hat{M}.(1)}{\hat{M}.(t)} \cdot E\left(\frac{m.(1)}{M.(1)}\right) - \tilde{\mu}_P(i;t) \quad .$$

Estimates of $E(m_i(1)/M_i(1))$ and $E(m.(1)/M.(1))$ are given in Apprendix A.

Analogously with state population, we estimate the covariance as follows:

$$\tilde{\sigma}_{CC}(i,j;t) = 0 \qquad t < 9$$

$$\tilde{\sigma}_{CC}(i,j;t) = -(.0008) \sum_k (\hat{C}_k(t))^2 / [\sum_{k \neq \ell} \hat{C}_k(t)\, \hat{C}_\ell(t)] \qquad i \neq j \quad t \geq 9 \quad .$$

§3.6 Per Capita Income (Substate level)

§3.6.1 Notation

Define

True (estimated) money income of county j in state i at time t	$M_{ij}(t)$ $(\hat{M}_{ij}(t))$
True (estimated) money income of local area k in county j in state i at time t	$M_{ijk}(t)$ $(\hat{M}_{ijk}(t))$
True (estimated) per capita income of county j in state i at time t	$C_{ij}(t)$ $(\hat{C}_{ij}(t))$
True (estimated) per capita income of local area k in county j in state i at time t	$C_{ijk}(t)$ $(\hat{C}_{ijk}(t))$.

Errors in estimates of money income and per capita income are denoted by $m_{ij}(t) = \hat{M}_{ij}(t) - M_{ij}(t)$, $m_{ijk}(t) = \hat{M}_{ijk}(t) - M_{ijk}(t)$, $c_{ij}(t) = \hat{C}_{ij}(t) - C_{ij}(t)$, $c_{ijk}(t) = \hat{C}_{ijk}(t) - C_{ijk}(t)$. The mean and variance of $c_{ij}(t)/C_{ij}(t) - c_i(t)/C_i(t)$ are denoted by $\mu_C(i,j;t)$ and $\sigma_C^2(i,j;t)$. However $\mu_C(i,j,k;t)$ and $\sigma_C^2(i,j,k;t)$ will denote the mean and variance of $c_{ijk}(t)/C_{ijk}(t)$ alone, and <u>not</u> the moments of $c_{ijk}(t)/C_{ijk}(t) - c_{ij}(t)/C_{ij}(t)$. The complicated structure of substate allocations (chapter 5) makes this non-uniform notation desirable. The covariance between $c_{ijk}(t)/C_{ijk}(t)$ and $c_{ij\ell}(t)/C_{ij\ell}(t)$ is denoted by $\sigma_{CC}(i,j,k,\ell;t)$.

Errors in estimates of net increase in money income and per capita income are denoted by

$$\Delta_M(i,j;t) = m_{ij}(t) - m_{ij}(1)$$

$$\Delta_M(i,j,k;t) = m_{ijk}(t) - m_{ijk}(1)$$

$$\Delta_C(i,j;t) = c_{ij}(t) - c_{ij}(1)$$

$$\Delta_C(i,j,k;t) = c_{ijk}(t) - c_{ijk}(1)$$

§3.6.2 Definitions and Data Sources

Total money income and per capita income are defined the same way at the substate level as at the state level. The reference periods for $C_{ij}(t)$, $C_{ijk}(t)$ are identical to those of $C_i(t)$ (see §3.5.2).

For EP's 1-5 the data sources for the substate per capita income estimates are the same as those at the state level. In EP 6 county per capita income estimates were updated by a procedure similar to that used to update the state-level estimates.

However, the EP 6 subcounty per capita income estimates were updated in a different manner. Essentially, the per capita estimates themselves were directly updated, rather than derived from separate updates of total money income and population (Herriot 1978; Office of Revenue Sharing 1976, pp.VI, VII). Subcounty (county) estimates were then adjusted to be consistent with county (state) estimates. Evaluations of the accuracy of the estimates have not been reported.

Estimates of per capita income for units with population under 500 were set equal to the estimate for the county in which the unit was situated.

Because $P_{ij}(t)$, $M_{ij}(t)$ and $P_{ijk}(t)$, $M_{ijk}(t)$ do not necessarily refer to the same year it is not in general true that $C_{ij}(t) = M_{ij}(t)/P_{ij}(t)$, $C_{ijk}(t) = M_{ijk}(t)/P_{ijk}(t)$. However, the following relations do hold

$$(3.25)\quad \begin{cases} C_{ij}(1) = M_{ij}(1)/P_{ij}(1) \;, & C_{ijk}(1) = M_{ijk}(1)/P_{ijk}(1) \;, \\ \hat{C}_{ij}(1) = \hat{M}_{ij}(1)/\hat{P}_{ij}(1) \;, & \hat{C}_{ijk}(1) = \hat{M}_{ijk}(1)/\hat{P}_{ijk}(1) \\ C_{ij}(9) = M_{ij}(9)/P_{ij}(9) \;, & C_{ijk}(9) = M_{ijk}(9)/P_{ijk}(9) \\ \hat{C}_{ij}(9) = \hat{M}_{ij}(9)/\hat{P}_{ij}(9) \;, & \hat{C}_{ijk}(9) = \hat{M}_{ijk}(9)/\hat{P}_{ijk}(9) \;. \end{cases}$$

Since $C_{ij}(t) = C_{ij}(1)$, $C_{ijk}(t) = C_{ijk}(1)$ for EP's 1-5 and $C_{ij}(t) = C_{ij}(9)$, $C_{ijk}(t) = C_{ijk}(9)$ for EP's 6-7, expression (3.25) is sufficient for decomposing $C_{ij}(t)$ and $C_{ijk}(t)$ for all EP's.

§3.6.3 Data Quality: Evidence

Major sources of error for substate estimates of total money income and per capita income include sampling variance, response variance, response bias, bias arising from population undercoverage, and updating error. SRI (1974a, p.III-56, Table 160.1) cites unpublished Census data to the effect that for $500 \leq P_{ij}(1)$, $P_{ijk}(1) \leq 2500$

$$\begin{aligned} &Var(\hat{C}_{ij}(t)/C_{ij}(t)) = 9/P_{ij}(t) \\ &Var(\hat{C}_{ijk}(t)/C_{ijk}(t)) = 9/P_{ijk}(t) \quad . \end{aligned} \qquad t = 1,\ldots 5$$

The accuracy of the updating procedure is unknown.

The study by Ono (1972, table 5, line 9) would seem to suggest that most of the error in $M_{ij}(1)$, $M_{ijk}(1)$ in New Jersey comes from net underreporting of social security income, public assistance income, and "other" income (defined in §3.5.2). The underreporting of social security income and public assistance income is believed to arise mainly from response bias and not undercoverage bias. In a statement before Congress (1973, p.2794) then Associate Director for Statistical Standards and Methodology of the U.S. Bureau of the Census, Joe Waksberg, stated:

> The reason why we missed most of these transfer payments, social security and welfare, is not because we are missing the people but when we ask questions on income, we do not get these kinds of income reported.

There is no evidence available about the relation between population undercoverage rates and the ratio of social security, public assistance, and "other" income to the total money income.

§3.6.4 Data Quality: Assumptions and Conclusions

The following proposition allows us to combine the evidence about different kinds of error in per capita income estimates.

Proposition 3.3 For $t = 1,9$ we have

$$(3.26)\quad \frac{c_{ij}(t)}{C_{ij}(t)} - \frac{c_i(t)}{C_i(t)} = \frac{\Delta_C(i,j;t)}{C_{ij}(t)} - \frac{\Delta_C(i;t)}{C_i(t)}$$

$$+ \frac{C_{ij}(1)}{C_{ij}(t)}\left(\frac{m_{ij}(1)}{M_{ij}(1)} - \frac{p_{ij}(1)}{P_{ij}(1)}\right) - \frac{C_i(1)}{C_i(t)}\left(\frac{m_i(1)}{M_i(1)} - \frac{p_i(1)}{P_i(1)}\right) + o_p(N^{-1/2})$$

and

$$(3.27)\quad \frac{c_{ijk}(t)}{C_{ijk}(t)} = \frac{\Delta_C(i,j,k;t)}{C_{ijk}(t)} + \frac{C_{ijk}(1)}{C_{ijk}(t)}\left(\frac{m_{ijk}(1)}{M_{ijk}(1)} - \frac{p_{ijk}(1)}{P_{ijk}(1)}\right) + o_p(N^{-1/2}) \quad .$$

Proof: First note the following identity

$$(3.28)\quad \frac{c_{ij}(t)}{C_{ij}(t)} - \frac{c_i(t)}{C_i(t)} = \frac{\Delta_C(i,j;t)}{C_{ij}(t)} - \frac{\Delta_C(i;t)}{C_i(t)}$$

$$+ \frac{C_{ij}(1)}{C_{ij}(t)}\,\frac{c_{ij}(1)}{C_{ij}(1)} - \frac{C_i(1)}{C_i(t)}\,\frac{c_i(1)}{C_i(1)} \quad .$$

By the delta method (theorem 2.1)

$$\frac{c_{ij}(1)}{C_{ij}(1)} = \frac{m_{ij}(1)}{M_{ij}(1)} - \frac{p_{ij}(1)}{P_{ij}(1)} + o_p(N^{-1/2})$$

and

$$\frac{c_i(1)}{C_i(1)} = \frac{m_i(1)}{M_i(1)} - \frac{p_i(1)}{P_i(1)} + o_p(N^{-1/2}) \quad .$$

Conclusion (3.26) follows upon substitution for $c_{ij}(1)/C_{ij}(1)$ and $c_i(1)/C_i(1)$ in (3.28). The proof of (3.27) is similar.

*****

Estimation of $\mu_C(i,j;t)$, $\sigma_C^2(i,j;t)$, $\mu_C(i,j,k;t)$ and $\sigma_C^2(i,j,k;t)$ can now proceed by estimation of means, variances, and covariances of terms on the right-hand side of (3.26) and (3.27).

Estimation of the moments of the right-hand side of (3.26) is facilitated by the approximate equality of $C_{ij}(1)/C_{ij}(t)$ and $C_i(1)/C_i(t)$ (e.g. for Essex County $\hat{C}_{ij}(1)/\hat{C}_{ij}(9) = .828$ while for New Jersey $\hat{C}_i(1)/\hat{C}_i(9) = .821$). Treating these ratios as equal simplifies the form of the estimates of the moments. Thus these moments will be approximated by those of

$$(3.29)\qquad \frac{\Delta_C(i,j;t)}{C_{ij}(t)} - \frac{\Delta_C(i;t)}{C_i(t)} + \frac{C_i(1)}{C_i(t)}\left[\frac{m_{ij}(1)}{M_{ij}(1)} - \frac{m_i(1)}{M_i(1)} - \left(\frac{p_{ij}(1)}{P_{ij}(1)} - \frac{p_i(1)}{P_i(1)}\right)\right]$$

The evidence of §3.6.3 suggests that the expected values of $m_{ij}(1)/M_{ij}(1)$ could vary over New Jersey counties because of variations in their total money income compositions with respect to social security income, public assistance income, and "other" income. Differential rates of population undercoverage are not that important. In principle the relative sizes of social security, public assistance, and "other" income could be determined for the various counties and estimates of $E(c_{ij}(1)/C_{ij}(1))$ formulated as for states in §3.5.4. This task is not undertaken in the present work and it is assumed instead that at the county level biases in money income are the same for each county.

$$(3.30)\qquad E\, m_{ij}(1)/M_{ij}(1) = E\, m_i(1)/M_i(1) \quad .$$

A consequence of the specification of $C_{ij}(t)$ is

$$\Delta_C(i;t) = \Delta_C(i,j;t) = \Delta_C(i,j,k;t) = 0 \qquad \text{for } t < 9 \quad .$$

The next assumption has the same basis as its state-level counterpart (3.21):

$$E\ [\Delta_M(i,j;t)/M_{ij}(t) - \Delta_M(i;t)/M_i(t)] = 0 \qquad t \geq 6 \quad . \tag{3.31}$$

It follows from (3.31), (3.22) and the kind of arguments used above that

$$E\ [\Delta_C(i,j;t)/C_{ij}(t) - \Delta_C(i;t)/C_i(t)] = 0 \qquad t \geq 6 \quad . \tag{3.32}$$

From (3.29), (3.30), and (3.32) the expectation of $c_{ij}(t)/C_{ij}(t) - c_i(t)/C_i(t)$ is approximately

$$-\frac{C_i(1)}{C_i(t)}\ \mu_P(i,j;t) \quad . \tag{3.33}$$

The evidence of §3.6.3 suggests that the variance of $c_{ij}(1)/C_{ij}(1)$ is $9/P_{ij}(1)$. Because $P_i(1)$ is so large, it is concluded that

$$\text{Var}\left(\frac{c_{ij}(1)}{C_{ij}(1)} - \frac{c_i(1)}{C_i(1)}\right) = 9/P_{ij}(1) \quad . \tag{3.34}$$

The variance of the county level updates is estimated as for states (§3.5.4). Thus

$$\tilde{\text{V}}\text{ar}\left(\frac{\Delta_C(i,j;9)}{C_{ij}(9)} - \frac{\Delta_C(i;9)}{C_i(9)}\right) \leq \left[\frac{\hat{C}_{ij}(9)-\hat{C}_{ij}(1)}{\hat{C}_{ij}(9)} - \frac{\hat{C}_i(9)-\hat{C}_i(1)}{\hat{C}_i(9)}\right]^2 \quad ,$$

and note that the average upper bound for counties in New Jersey was 0.0001. Using this upper bound as a conservative estimate (it was argued in §3.5.4 that this kind of "conservative" estimate might in fact be rather accurate) we derive*

$$(3.35) \qquad \mathrm{Var}\left[\frac{\Delta_C(i,j;9)}{C_{ij}(9)} - \frac{\Delta_C(i;9)}{C_i(9)}\right] = .0001 \ .$$

For subcounty areas the same kinds of arguments and assumptions yield

$$(3.36) \qquad \mathrm{Var}(c(i,j,k;1)/C_{ijk}(1)) = 9/P_{ijk}(1)$$

and

$$(3.37) \qquad \mathrm{Var}(\ \Delta_C(i,j,k;9)/C_{ijk}(9)) = .0006 \ .$$

The bias terms are also derived similarly but appear different than those for counties because $\mu_C(i,j,k;t)$ is not a differential bias. Thus

$$(3.38) \qquad E\ (c_{ijk}(1)/C_{ijk}(1)) = \frac{C_{ijk}(1)}{C_{ijk}(t)} \frac{P_{ijk}(1)}{P_{ijk}(t)} \tilde{A}_{ijk}$$

and

$$(3.39) \qquad E\ (\Delta_C(i,j,k;9)/C_{ijk}(9)) = \text{constant} \ ,$$

---

* A recent paper by Fay and Herriot (1979) suggests that the variances in (3.35) and (3.37) may be far greater than .0001 and .0006.

where the constant term corresponds to $E(\Delta_C(i,j;9)/C_{ij}(9))$. In applications we will only be concerned with

$$E(\frac{c_{ijk}(t)}{C_{ijk}(t)} - \sum_{\ell} W_\ell \frac{c_{ij\ell}(t)}{C_{ij\ell}(t)} \quad \text{where} \quad \sum_{\ell} W_\ell = 1$$

and the constant term will not matter.

Positive correlation between $c_{ijk}(t)/C_{ijk}(t)$ and $c_{ij\ell}(t)/C_{ij\ell}(t)$ occurs because (i) the same updating methodology was applied to all subcounty per capita income estimates, and (ii) the subcounty estimates were all adjusted for consistency with the county updates. Since per capita income estimates were not updated for $t < 9$, the errors arise solely from sampling variance and response variance and are uncorrelated. Therefore

$$\sigma_{CC}(i,j,k,\ell;t) = \text{Cov}(c_{ijk}(t)/C_{ijk}(t) \ , \ c_{ij\ell}(t)/C_{ij\ell}(t)) = 0 \qquad t < 9 \ .$$

For $t \geq 9$ the Cauchy-Schwartz inequality implies

$$\sigma_{CC}(i,j,k,\ell;t) \leq (\text{Var}\Delta_{ijk}(t)/C_{ijk}(t))^{1/2} \ (\text{Var}\Delta_{ij\ell}(t)/C_{ij\ell}(t))^{1/2}$$

and so from (3.37) and the non-negativity of the correlation:

$$0 \leq \tilde{\sigma}_{CC}(i,j,k,\ell;t) \leq .0006 \qquad t \geq 9 \ .$$

The remainder of this subsection presents arguments that

(3.40) $$\text{Cov}(\frac{c_{ij}(t)}{C_{ij}(t)} - \frac{c_i(t)}{C_i(t)} \ , \ \frac{c_i(t)}{C_i(t)} - \frac{c(t)}{C(t)})$$

and

(3.41) $$\text{Cov}(\frac{c_{ijk}(t)}{C_{ijk}(t)} - \frac{c_{ij}(t)}{C_{ij}(t)} \ , \ \frac{c_{ij}(t)}{C_{ij}(t)} - \frac{c_i(t)}{C_i(t)})$$

are negligible.

First note that empirically $C_{ij}(1)/C_{ij}(t)$, $C_i(1)/C_i(t)$, and $C(1)/C(t)$ are close (for i = New Jersey, j = Essex County and t = 9, estimates of the ratios based on GRS data were .828, .821, .825 respectively). Using identities like (3.28) it follows that (3.40) is approximately equal to

$$\mathrm{Cov}\left(\frac{\Delta_C(i,j;t)}{C_{ij}(t)} - \frac{\Delta_C(i;t)}{C_i(t)} + \gamma\left(\frac{c_{ij}(1)}{C_{ij}(1)} - \frac{c_i(1)}{C_i(1)}\right),\right.$$

$$\left.\frac{\Delta_C(i;t)}{C_i(t)} - \frac{\Delta_C(t)}{C(t)} + \gamma\left(\frac{c_i(1)}{C_i(1)} - \frac{c(1)}{C(1)}\right)\right)$$

where $\gamma = c(1)/C(t)$. It is believed that random errors in census estimates of per capita income (arising from response variance) are uncorrelated with errors in the updates (arising largely from errors in methodology and administrative records). To show that (3.40) is small it therefore suffices to show that each of

$$\text{(3.42)} \qquad \mathrm{Cov}\left(\frac{\Delta_C(i,j;t)}{C_{ij}(t)} - \frac{\Delta_C(i;t)}{C_i(t)}, \frac{\Delta_C(i;t)}{C_i(t)} - \frac{\Delta_C(t)}{C(t)}\right)$$

and

$$\text{(3.43)} \qquad \mathrm{Cov}\left(\frac{c_{ij}(1)}{C_{ij}(1)} - \frac{c_i(1)}{C_i(1)}, \frac{c_i(1)}{C_i(1)} - \frac{c(1)}{C(1)}\right)$$

is small.

To see that (3.43) is small consider the following lemma.

Lemma 3.4 For $\ell = 1,\ldots N_{ij}$ let $X_{ij\ell}$ be uncorrelated random variables with $\mathrm{Var}(X_{ij\ell}) = \lambda C_{ij}^2$ and let $C_i = \sum_j N_{ij}C_{ij}/N_{i\cdot}$, and $C = \sum_i (N_{i\cdot}C_i/N_{\cdot\cdot})$. Then

$$\mathrm{Cov}\left(\frac{X_{ij\cdot}}{C_{ij}N_{ij}} - \frac{X_{i\cdot\cdot}}{C_iN_i}, \frac{X_{i\cdot\cdot}}{C_iN_{i\cdot\cdot}} - \frac{X_{\cdot\cdot\cdot}}{C\,N_{\cdot\cdot}}\right)$$

$$= \lambda\left(\frac{1}{C_iN_{i\cdot}} - \frac{1}{C\,N_{\cdot\cdot}}\right)\left(C_{ij} - \frac{\sum_{j'} N_{ij'}C_{ij'}^2}{C_iN_{i\cdot}}\right)$$

Proof: Simply observe that

$$\mathrm{Cov}\left(\frac{X_{ij\cdot}}{C_{ij}N_{ij}} - \frac{X_{i\cdot\cdot}}{C_iN_{i\cdot}}, \frac{X_{i\cdot\cdot}}{C_iN_{i\cdot}} - \frac{X_{\cdot\cdot\cdot}}{C\,N_{\cdot\cdot}}\right)$$

$$= \mathrm{Cov}\left(\frac{X_{ij\cdot}}{C_{ij}N_{ij}}, \frac{X_{i\cdot\cdot}}{C_iN_{i\cdot}}\right) - \mathrm{Cov}\left(\frac{X_{ij\cdot}}{C_{ij}N_{ij}}, \frac{X_{\cdot\cdot\cdot}}{C\,N_{\cdot\cdot}}\right)$$

$$- \mathrm{Var}\left(\frac{X_{i\cdot\cdot}}{C_iN_{i\cdot}}\right) + \mathrm{Cov}\left(\frac{X_{i\cdot\cdot}}{C_iN_{i\cdot}}, \frac{X_{\cdot\cdot\cdot}}{C\,N_{\cdot\cdot}}\right)$$

$$= \frac{N_{ij}\,\lambda C_{ij}^2}{C_{ij}N_{ij}C_iN_{i\cdot}} - \frac{N_{ij}\,\lambda C_{ij}^2}{C_{ij}N_{ij}C\,N_{\cdot\cdot}}$$

$$- \frac{\lambda\sum_{j'} N_{ij'}C_{ij'}^2}{(C_iN_{i\cdot})^2} + \frac{\lambda\sum_{j'} N_{ij'}C_{ij'}^2}{(C_iN_{i\cdot})C\,N_{\cdot\cdot}}$$

$$= \lambda\left(\frac{1}{C_iN_{i\cdot}} - \frac{1}{C\,N_{\cdot\cdot}}\right)\left(C_{ij} - \frac{\sum_{j'} N_{ij'}C_{ij'}^2}{C_iN_{i\cdot}}\right).$$

*****

To use the lemma let $X_{ij\ell}$ denote the reported income of individual ℓ in county j, state i and let N_{ij} be the sample size for county j, so N_{ij} is about $.20\ P_{ij}(1)$. Here C_{ij}, C_i denote $C_{ij}(1)$, $C_i(1)$ and $\lambda C^2_{ij}(1)$ is the variance of X_{ij} (where $\lambda = (.20)9 = 1.8$). Because $C_{ij}(1)$ was relatively constant the covariance is small. Calculations on EP 1 New Jersey data give

$$\frac{\sum_j N_{ij} C_{ij}^{\ 2}}{C_i N_{i\cdot}} = 3732 \quad .$$

The county for which $\hat{C}_{ij}(1)$ differed most from 3732 was Cumberland County, with $\hat{C}_{ij}(1) = 2882$. Since $\hat{C}_i(1) = 3674$ it follows that the absolute value of the covariance is less than

$$\lambda\left(\frac{1}{.2P_i(1)\ 3674}\right)(3732 - 2882)$$

$$< \frac{1.2\lambda}{P_i(1)} \quad .$$

Since $1.2\lambda/P_i(1)$ is so much smaller than $\lambda/P_{ij}(1)$, the covariance (3.43) is negligible compared to the variance in EP 1, $\lambda/P_{ij}(1)$. For example $\hat{P}_i(1) = 7.17$ million while for Cumberland County $\hat{P}_{ij}(1) = 0.12$ million.

Because county per capita income updates were adjusted to be consistent with state-level updates, lemma 3.2 can be used to show that (3.42) is $o(N^{-1})$. Since (3.42), (3.43) are both small, (3.40) must be small. The proof for (3.41) is similar.

§3.6.5 Relevant Distributions and Approximations

From (3.26), (3.33)-(3.35) and the assumptions in the previous section the mean and variance of $c_{ij}(t)/C_{ij}(t) - c_i(t)/C_i(t)$ are estimated by

$$\tilde{\mu}_C(i,j;t) = -\frac{\hat{c}_i(1)}{C_i(t)}\,\tilde{\mu}_P(i,j;t)$$

and

$$\tilde{\sigma}_C^2(i,j;t) = 9/\hat{P}_{ij}(1) \qquad t < 9$$

$$= \frac{\hat{C}_i(1)^2}{\hat{C}_i(t)} \cdot \frac{9}{\hat{P}_{ij}(1)} + .0001 \qquad t \geq 9 \quad .$$

Using (3.27), (3.36)-(3.38) and associated assumptions, the mean and variance of $c_{ijk}(t)/C_{ijk}(t)$ will be estimated by

$$\tilde{\mu}_C(i,j,k;t) = \frac{\hat{C}_{ijk}(1)}{\hat{C}_{ijk}(t)}\,\frac{\hat{P}_{ijk}(1)}{\hat{P}_{ijk}(t)}\,\tilde{A}_{ijk} + \text{constant term}$$

and

$$\tilde{\sigma}_C^2(i,j,k;t) = 9/\hat{P}_{ijk}(1) \qquad t < 9$$

$$= \frac{\hat{C}_{ijk}(1)}{\hat{C}_{ijk}(t)}^{2} \cdot \frac{9}{\hat{P}_{ijk}(1)} + .0006 \qquad t \geq 9 \quad ,$$

where the constant term is unknown but unimportant for the present applications (recall discussion in the preceding section). Earlier discussion (§3.6.4) also suggests that $\sigma_{CC}(i,j,k,\ell;t)$, the covariance between $c_{ijk}(t)/C_{ijk}(t)$ and $c_{ij\ell}(t)/C_{ij\ell}(t)$, satisfies

$$\sigma_{CC}(i,j,k,\ell;t) = 0 \qquad t > 9$$

and

$$0 \leq \sigma_{CC}(i,k,k,\ell;t) \leq .0006 \qquad t \geq 9 \quad .$$

The estimates $\tilde{\mu}_C(i,j;t)$, $\tilde{\mu}_C(i,j,k;t)$ appear to be naive in that they neglect response bias and overstate the effect of population undercoverage. Although they may fail to be reliable estimates of individual bias rates, these estimates are useful because on the whole they may be indicative of the variation in the bias rates over ensembles of substate units.

§3.7 Personal Income

§3.7.1 Notation

Denote the true (estimated) personal income of state i at time t by $R_i(t)$ $(\hat{R}_i(t))$. Denote the error in the GRS estimate by $r_i(t) = \hat{R}_i(t) - R_i(t)$ with estimate $\tilde{r}_i(t)$. The mean and variance of $r_i(t)/R_i(t) - r.(t)/R.(t)$ are written $\mu_R(i;t)$ and $\sigma_R^2(i;t)$ with estimates $\tilde{\mu}_R(i;t)$, $\tilde{\sigma}_R^2(i;t)$. The covariance of $r_i(t)/R_i(t) - r.(t)/R.(t)$ with $r_j(t)/R_j(t) - r.(t)/R.(t)$ is denoted by $\sigma_{RR}(i,j;t)$, with estimate $\tilde{\sigma}_{RR}(i,j;t)$.

§3.7.2 Definition and Data Sources

Personal income is a statistical construct of the Bureau of Economic Analysis. It is designed to measure the income of individuals for national income account purposes (Coleman 1974; SRI 1974a, p.III-12 ff). Personal income and money income are not strictly comparable because the definitions include distinct sources of income and because the personal income estimates include income of some individuals not included in estimates of money income; see Census (1975b No.575, pp.IV,V). The procedure used to estimate personal income is very complex, drawing upon several hundred data sources. The bulk of the data comes from administrative records of various federal and state government programs. The remainder comes from various censuses, surveys, and non-government sources (Coleman 1977).

The following specification holds.

t	EP	$R_i(t)$ refers to calendar year
1,2	1,2	1970
3,5	3,4	1970
7	5	1971
9	6	1972
10	7	1973

Table 3.4

Target Variables - Personal Income

§3.7.3 Data Quality: Evidence

Following Morgenstern (1973, chapter 14, sec.3) we distinguish three principle sources of error in estimates of personal income: errors in basic data, gaps in basic data, and errors resulting from forcing available statistics into the conceptual framework of personal income. Examples of the latter are errors from converting data from a fiscal to calendar year basis.

Quantitative estimates of error are not published and it is difficult to know the magnitude of error. One source of error that can be measured is "revision error", which arises because the personal income estimates are not based on the latest revised data (SRI 1974a, pp. III-13, III-14). In EP's 1-4 the average absolute deviation of the estimates $\hat{R}_i(t)$ from the revised estimates was .0148 times $\hat{R}_i(t)$ (ibid).

§3.7.4 Data Quality: Assumptions and Conclusions

Assuming (from central limit theory considerations) that the revision errors are normally distributed, we can estimate the contribution of revision error variance to $\sigma_R^2(i;t)$ at $(\pi/2)(.0148)^2$.

How much of the total error $r_i(t)$ is captured by the revision error is not known. For a ballpark estimate we will use $3(\pi/2)(.0148)^2$ to estimate $\sigma_R^2(i;t)$.

The available evidence does not suggest any significant biases in $R_i(t)$ so we estimate

$$\mu_R(i;t) = E(\frac{\hat{R}_i(t) - R_i(t)}{R_i(t)} - \frac{\hat{R}.(t) - R.(t)}{R.(t)})$$

by $\tilde{\mu}_R(i;t) = 0$.

§3.7.5 Relevant Distributions and Approximations

The error model used will be

$$\tilde{\mu}_R(i;t) = 0$$

$$\tilde{\sigma}_R^2(i;t) = 10.32 \times 10^{-4} .$$

Analogously with state population, the covariance is estimated to be

$$\tilde{\sigma}_{RR}(i,j;t) = -(10.32 \times 10^{-4}) \sum_k (\hat{R}_k(t))^2 / \sum_{k \neq \ell} \hat{R}_k(t)\hat{R}_\ell(t) \quad i \neq j .$$

§3.8 Net State and Local Taxes

§3.8.1 Notation

Denote the true (estimated) net state and local taxes of state i at time t by $T_i(t)$ $(\hat{T}_i(t))$. The error in the GRS estimate will be denoted by $t_i(t) = \hat{T}_i(t) - T_i(t)$, with estimate $\tilde{t}_i(t)$. The mean and variance of $t_i(t)/T_i(t) - t.(t)/T.(t)$ will be written $\mu_T(i;t)$ and $\sigma_T^2(i;t)$, with estimates $\tilde{\mu}_T(i;t)$, $\tilde{\sigma}_T^2(i;t)$. The covariance of $t_i(t)/T_i(t) - t.(t)/T.(t)$ with $t_j(t)/T_j(t) - t.(t)/T.(t)$ is denoted by $\sigma_{TT}^2(i,j;t)$, with estimate $\tilde{\sigma}_{TT}^2(i,j;t)$.

§3.8.2 Definition and Data Source

The net state and local taxes of a state are (Office of Revenue Sharing 1976, EP 6, pp.xiv,xv):

> the compulsory contributions exacted by the State (or by any unit of local government or other political subdivision of the State) for public purposes (other than employee and employer assessments and contributions to finance retirement and social insurance systems, and other than special assessments for capital outlay) as such contributions are determined by the Bureau of Census for general statistical purposes.

Thus $T_i(t)$ is the sum of state and local taxes.

The following specification holds

t	EP	$T_i(t)$ refers to fiscal year
1,2	1,2	1970-1971
3,5	3,4	1970-1971
7	5	1971-1972
9	6	1972-1973
10	7	1973-1974

Table 3.5

Target Variables - Net State and Local Taxes

The estimates are published in Census (1972b,No.5; 1973d, No.5; 1974d, No.5; 1975c, No.5). The data is obtained as follows (Office of Revenue Sharing 1976, EP 6, p.xv):

> The State government information contained in State and local taxes is based on the annual Bureau of the Census survey of State finances. State finances statistics are compiled by representatives of the Bureau of the Census from official records and reports of the various States. The local government portion of the State and local taxes data are estimates based on information received from a sample of such governments. The sample consisted of approximately 16,000 local governments and included all counties having a 1970 population of 50,000 or more, all cities having a 1970 population of 25,000 or more, all other governments whose relative importance in their State based on expenditure or debt was above a specified size, and a random sample of the remaining units.

§3.8.3 Data Quality: Evidence

According to the Census Bureau (Census 1976b, Table E), the relative sampling standard error in fiscal year 1974-75 property tax data (a component of net state and local taxes) was less than 1% for 28 states, between 1% and 1.9% for 21 states, and between 2% and 2.9% for two states. Extensive editing and auditing is performed to control response error and data processing error (SRI 1974a, pp.III-3, III-33).

§3.8.4 Data Quality: Assumptions and Conclusions

It is assumed that the sampling error rate of property tax data is indicative of the sampling error rates of other components of net state and local taxes. Response errors are assumed negligible. Because $\hat{T}_i(t)$ and $T_i(t)$ refer to the same fiscal years, errors from timeliness are also negligible. A reasonable estimate of $\sigma_T^2(i;t)$ is

$$(.013)^2 = [(.01)^2(28) + (.015)^2(21) + (.025)^2(2)]/51 \quad .$$

§3.8.5 Relevant Distributions and Approximations

The error model is

$$\tilde{\mu}_T(i;t) = 0$$

and

$$\tilde{\sigma}_T^2(i;t) = (0.013)^2 \quad .$$

Analogously with state population, the covariance is estimated by

$$\tilde{\sigma}_{TT}(i,j;t) = -(0.013)^2 \sum_k (T_k(t))^2 / \sum_{k \neq \ell}\sum T_k(t)T_\ell(t) \qquad i \neq j \quad .$$

§3.9 State Individual Income Taxes

§3.9.1 Notation

Denote the true (estimated) state individual income tax collection for state i at time t by $K_i(t)$ $(\hat{K}_i(t))$. The error in the GRS estimate is denoted by $k_i(t) = \hat{K}_i(t) - K_i(t)$, with the estimate $\tilde{k}_i(t)$. The mean and variance of $k_i(t)/K_i(t)$ are denoted by $\mu_K(i;t)$ and $\sigma_K^2(i;t)$, with estimates $\tilde{\mu}_K(i;t)$, $\tilde{\sigma}_K^2(i;t)$. Notice that μ_K and σ_K^2 are moments of $k_i(t)/K_i(t)$ and <u>not</u> of $k_i(t)/K_i(t) - k.(t)/K.(t)$.

§3.9.2 Definition and Data Sources

State individual income taxes are (Office of Revenue Sharing 1976, p.xiii):

> the total calendar year...collections of the tax imposed upon the income of individuals by such State and described as State income tax under section 164 (a) (3) of the I.R.S. code of 1954. These are collections of taxes on individuals measured by net income and taxes on special types of income (e.g. interest, dividends, income from intangibles, etc.).

Estimates for EP's 1 and 2 were calculated by averaging the 1972 and 1973 fiscal year predictions made by the Fiscal Officer of each state in response to a survey taken in the first two months of 1972 by the Joint Committee on Internal Revenue Taxation. For later EP's the estimates $\hat{K}_i(t)$ were based on state records of the amount of tax collected. Preliminary data may be found in <u>Quarterly Summary of State and Local Tax Revenue</u> (Census 1972c). The estimates $\hat{K}_i(t)$ incorporate revisions to these data; see Census (1972b No. 3; 1973d No. 3; 1974d No. 3; 1975c No. 3).

Note that $K_i(t)$ refers to tax collections, not tax liabilities, net of refunds.

The following specification holds.

t	EP	$K_i(t)$ refers to calendar year
1,2	1,2	1972
3,5	3,4	1972
7	5	1973
9	6	1974
10	7	1975

Table 3.6

Target Variables - State Individual Income Taxes

§3.9.3 Data Quality: Evidence

The estimates for EP's 1 and 2 were subject to errors from the prediction process. Sources of error for $\hat{K}_i(t)$ $t > 2$ include late payments, refunds, and other adjustments. The error in $\hat{K}_i$ for EP's 3-7 is thus much smaller than the errors for EP's 1 and 2. In particular, comparison of $\hat{K}_i(1)$ or $\hat{K}_i(2)$ (they were the same) with $\hat{K}_i(3)$ provides evidence concerning the magnitude of the EP 1,2 errors. For the 45 states collecting individual income taxes, the average value of $(\hat{K}_i(2) - \hat{K}_i(3))/\hat{K}_i(3)$ was -0.071, the average squared value was 0.0174, and the sample variance was 0.0127.

§3.9.4 Data Quality: Assumptions and Conclusions

If we assume that the relative errors $k_i(t)/K_i(t)$ $t = 1,2$ are identically distributed, the evidence above suggests that (for states collecting individual income tax)

$$\mu_K(i;t) = E\,(k_i(t)/K_i(t)) = -0.071 \qquad t = 1,2$$

and

$$\sigma_K^2(i;t) = \mathrm{Var}(k_i(t)/K_i(t)) = 0.0127 \qquad t = 1,2 \quad .$$

It is assumed that the expected ratio of the sum of tax adjustments and late payments (after the estimate $\hat{K}_i(t)$ has been formed) to $K_i(t)$ is zero for all states. This assumption could be tested by laborious examination of states' fiscal accounts. We have therefore

$$\tilde{\mu}_K(i;t) = 0 \qquad t > 2 \quad .$$

The large number of state individual income tax payers in each state makes it plausible that for $t > 2$ $\sigma_K^2(i;t)$ is negligible.

3.9.5 Relevant Distributions and Approximations

For states collecting individual income tax, we have

$$\tilde{\mu}_K(i;t) = 0.071 \qquad t = 1,2$$

$$= 0.0 \qquad t > 2$$

and

$$\tilde{\sigma}_K^2(i;t) = 0.0127 \qquad t = 1,2$$

$$= 0.0 \qquad t > 2 \quad .$$

§3.10 Federal Individual Income Tax Liabilities

§3.10.1 Notation

Denote the true (estimated) federal individual income tax liability of state i at time t by $L_i(t)$ $(\hat{L}_i(t))$. Let $l_i(t) = \hat{L}_i(t) - L_i(t)$, denote the mean and variance of $l_i(t)/L_i(t)$ by $\mu_L(i;t)$ and $\sigma_L^2(i;t)$, and denote estimates of these quantities by $\tilde{l}_i(t)$, $\tilde{\mu}_L(i;t)$, $\tilde{\sigma}_L^2(i;t)$. Note that $\mu_L(i;t)$, $\sigma_L^2(i;t)$ are moments of $l_i(t)/L_i(t)$, not of $l_i(t)/L_i(t) - l.(t)/L.(t)$.

§3.10.2 Definition and Data Sources

According to Office of Revenue Sharing (1976, p.xiii)

> The Federal individual income tax liability of a State for revenue sharing purposes is the total annual Federal individual income taxes after credits attributed to the residents of the State by the Internal Revenue Service. Income tax after credits is determined by subtracting statutory credits from the total of income tax before credits and tax surcharge.

The following specification holds.

t	EP	$L_i(t)$ refers to calendar year
1,2	1,2	1971
3,5	3,4	1971
7	5	1972
9	6	1973
10	7	1974

Table 3.7

Target Variables - Federal Individual Income Tax Liabilities

The estimation is based on a stratified sample of over 250,000 unaudited tax returns.

§3.10.3 Data Quality: Evidence

The sampling procedure is described in Internal Revenue Service (1973, pp.316-317). Individual estimates of $\mu_L(i;t)$, $\sigma_L^2(i;t)$ for states are not available, although sampling theory provides for straightforward estimates of these parameters. At the national level the coefficient of variation for income tax after credits was estimated to be 0.2% for 1971 (ibid, p.318, table 7c, col.92, line 1). Considerable checking and controls are maintained.

§3.10.4 Data Quality: Assumptions and Conclusions

Estimates of $\sigma_L^2(i;t)$ will be derived from the assumptions that (1) $\sigma_L^2(i;t) = \text{Var}(l_i(t)/L_i(t))$ is constant over all states and (2) the $\hat{L}_i(t)$ are uncorrelated. Notice

$$\text{Var}(l.(t)/L.(t)) = \text{Var}(\Sigma \frac{L_i(t)}{L.(t)} \frac{l_i(t)}{L_i(t)})$$

$$= \Sigma \left(\frac{L_i(t)}{L.(t)}\right)^2 \cdot \sigma_L^2(i;t)$$

and since $\sigma_L^2(i;t)$ is constant for all i,

$$\sigma_L^2(i;t) = \text{Var}(l.(t)/L.(t))/\Sigma \left(\frac{L_i(t)}{L.(t)}\right)^2 .$$

Using GRS data to estimate $\Sigma\left(\frac{L_i(t)}{L.(t)}\right)^2$ and $(.002)^2$ to estimate $\text{Var}(l.(t)/L.(t))$ gives

$$\tilde{\sigma}_L^2(i;t) = (.002)^2/\Sigma \left(\frac{\hat{L}_i(t)}{\hat{L}.(t)}\right)^2$$

The available evidence does not suggest any bias in $\hat{L}_i(t)/L_i(t)$, so $\mu_L(i;t)$ is estimated to be zero.

§3.10.5 Relevant Distribution and Approximation

The error model is

$$\tilde{\mu}_L(i;t) = 0$$

$$\tilde{\sigma}_L^2(i;t) = (.002)^2/[\Sigma\ (\hat{L}_i(t)/\hat{L}.(t))^2] \quad .$$

§3.11 Income Tax Amount (State level)

§3.11.1 Notation

In this section the subscript j will denote state j . Define the true (estimated) income tax amount of state j :

$$I_j(t) = \text{median}\ \{.01L_j(t)\ ,\ .15K_j(t)\ ,\ .06L_j(t)\}$$

$$(\hat{I}_j(t) = \text{median}\ \ \{.01\hat{L}_j(t)\ ,\ .15\hat{K}_j(t)\ ,\ .06\hat{L}_j(t)\}\ ) \quad .$$

Errors in GRS estimates of income tax amount are denoted by $i_j(t) = \hat{I}_j(t) - I_j(t)$, $i.(t) = \hat{I}.(t) - I.(t)$ with estimates $\tilde{i}_j(t)$, $\tilde{i}.(t)$. The mean and variance of $i_j(t)/I_j(t) - i.(t)/I.(t)$ are denoted by $\mu_I(j;t)$, $\sigma_I^2(j;t)$ with estimates $\tilde{\mu}_I(j;t)$, $\tilde{\sigma}_I^2(j;t)$. The covariance of $i_j(t)/I_j(t) - i.(t)/I.(t)$ with $i_{j'}(t)/I_{j'}(t) - i.(t)/I.(t)$ is denoted by $\sigma_{II}(j,j';t)$ with estimates $\tilde{\sigma}_{II}(j,j';t)$. We also define the sets of indices

$$\mathcal{K}(t) = \{j: I_j(t) = .15K_j(t)\}$$

$$\mathcal{L}(t) = \{j: I_j(t) = .01L_j(t) \text{ or } I_j(t) = .06L_j(t)\}$$

$$\hat{\mathcal{K}}(t) = \{j: \hat{I}_j(t) = .15\hat{K}_j(t)\}$$

$$\hat{\mathcal{L}}(t) = \{j: \hat{I}_j(t) = .01\hat{L}_j(t) \text{ or } \hat{I}_j(t) = .06\hat{L}_j(t)\} .$$

§3.11.2 Definition and Data Sources

According to the legislation (Public Law 92-512, Title I, sec. 109(b)(2)-(3))

> The income tax amount of any State...is 15 per cent of the net amount collected from the State individual income tax...The income tax amount...(a) shall not exceed 6 per cent, and (b) shall not be less than 1 per cent, of the Federal individual income tax liabilities attributed to such State...

There are no new data sources to consider since $K_i(t)$, $L_i(t)$ have been discussed in §3.9 and §3.10.

§3.11.3 Data Quality: Evidence

The quality of the $\hat{I}_j$ estimates depends upon the quality of the $\hat{L}_j$, $\hat{K}_j$ estimates and upon the composition of the sets $\mathcal{L}(t)$, $\mathcal{K}(t)$. In EP's 1 and 2 the set $\hat{\mathcal{L}}(t)$ included those states with no state individual income tax (Florida, Nevada, South Dakota, Texas, Washington, Wyoming) and also Connecticut, Minnesota, New Hampshire, New Jersey, and Tennessee.

§3.11.4 Data Quality: Assumption and Conclusion

Recall from §3.9.5 and §3.10.5 that for $t = 1,2$ we estimated

$$\tilde{\mu}_K(j;t) = E(k_j(t)/K_j(t)) \text{ by } -0.071$$

and

$$\tilde{\mu}_L(j;t) = E(l_j(t)/L_j(t)) \text{ by } 0.0 \quad .$$

Accordingly, for $t = 1,2$ we estimate $E(i.(t)/I.(t)$ by

$$(-0.071) \Big(\sum_{j \varepsilon \mathcal{K}(t)} \hat{I}_j(t)/\hat{I}.(t) \Big) = -0.063 \quad .$$

Since $\tilde{\mu}_L(i;t) = \tilde{\mu}_K(i;t) = 0.0$ for $t > 2$, the estimate of $E(i.(t)/I.(t))$ is 0.0 for $t > 2$.

For $j \varepsilon \hat{\mathcal{K}}(t)$ we estimate $\mu_I(j;t)$ by using $\mu_I(j;t) = \mu_K(j;t) - E(i.(t)/I.(t))$ and analogously for $j \varepsilon \hat{\mathcal{L}}(t)$ we use

$$\mu_I(j;t) = \mu_L(j;t) - E(i.(t)/I.(t)) \quad .$$

This yields the estimates

$$\begin{aligned} \mu_I(j;t) &= -0.008 \qquad t = 1,2 \qquad j \varepsilon \hat{\mathcal{K}}(t) \\ &= +0.063 \qquad t = 1,2 \qquad j \varepsilon \hat{\mathcal{L}}(t) \\ &= \ \ 0.0 \qquad\ \ t > 2 \quad . \end{aligned}$$

To estimate $\sigma_I^2(i;t)$ first notice that

$$\mathrm{Var}(i.(t)/I.(t)) = \Sigma\, (I_j(t)/I.(t))^2\ \mathrm{Var}(i_j(t)/I_j(t))$$

and

$$\mathrm{Cov}(i_j(t)/I_j(t)\ ,\ i.(t)/I.(t)) = (I_j(t)/I.(t))\ \mathrm{Var}(i_j(t)/I_j(t))\ .$$

Consequently we have

$$\begin{aligned}\sigma_I^2(j;t) &= \mathrm{Var}(i_j(t)/I_j(t) - i.(t)/I.(t)) \\ &= \mathrm{Var}(i_j(t)/I_j(t)) + \mathrm{Var}(i.(t)/I.(t)) \\ &\quad - 2\ \mathrm{Cov}(i_j(t)/I_j(t)\ ,\ i.(t)/I.(t)) \\ &= (1 - 2I_j(t)/I.(t))\ \mathrm{Var}(i_j(t)/I_j(t)) \\ &\quad + \sum_{j'} (I_{j'}(t)/I.(t))^2\ \mathrm{Var}(i_{j'}(t)/I_{j'}(t))\ .\end{aligned}$$

For $j \varepsilon\ \hat{K}(t)\ ,\ \hat{L}(t)$ respectively $\mathrm{Var}(i_j(t)/I_j(t))$ is estimated by $\tilde{\sigma}_K^2(j;t)\ ,\ \tilde{\sigma}_L^2(j;t)$, so $\sigma_I^2(j;t)$ is estimated by

$$(3.44)\qquad \sum_{j' \varepsilon \hat{K}(t)} (\hat{I}_{j'}(t)/\hat{I}.(t))^2\ \tilde{\sigma}_K^2(j';t) + \sum_{j' \varepsilon \hat{L}(t)} (\hat{I}_{j'}(t)/\hat{I}.(t))^2\ \tilde{\sigma}_L^2(j';t) + \left(1-2\,\frac{\hat{I}_j(t)}{\hat{I}.(t)}\right) \cdot \begin{cases} \tilde{\sigma}_K^2(j;t) & j \varepsilon\ \hat{K}(t) \\ \tilde{\sigma}_K^2(j;t) & j \varepsilon\ \hat{L}(t) \end{cases}$$

§3.11.5 Relevant Distributions and Approximations

From the preceding section we estimate

$$\tilde{\mu}_I(j;t) = -0.008 \qquad j \varepsilon \hat{K}(1) \qquad t = 1,2$$
$$\tilde{\mu}_I(j;t) = +0.063 \qquad j \varepsilon \hat{L}(1) \qquad t = 1,2$$
$$\tilde{\mu}_I(j;t) = 0 \qquad\qquad t > 2 \quad .$$

Estimates $\tilde{\sigma}_I^2(j;t)$ are obtained from (3.44).

§3.12 Adjusted Taxes (Substate level)

§3.12.1 Notation

Define

True (estimated) adjusted taxes of county government j in state i at time t — $D_{ij}(t)$ $(\hat{D}_{ij}(t))$

True (estimated) adjusted taxes of subcounty government k in county j in state i at time t — $D_{ijk}(t)$ $(\hat{D}_{ijk}(t))$.

Errors in GRS estimates of adjusted taxes are denoted by $d_{ij}(t) = \hat{D}_{ij}(t) - D_{ij}(t)$ and $d_{ijk}(t) = \hat{D}_{ijk}(t) - D_{ijk}(t)$ with estimates $\tilde{d}_{ij}(t)$, $\tilde{d}_{ijk}(t)$. The means and variances of $d_{ij}(t)/D_{ij}(t)$ and $d_{ijk}(t)/D_{ijk}(t)$ are denoted by $\mu_D(i,j;t)$, $\sigma_D^2(i,j;t)$, $\mu_D(i,j,k;t)$ and $\sigma_D^2(i,j,k;t)$, with estimates $\tilde{\mu}_D(i,j;t)$, $\tilde{\sigma}_D^2(i,j;t)$, $\tilde{\mu}_D(i,j,k;t)$ and $\tilde{\sigma}_D^2(i,j,k;t)$.

Note that $D_{ij}(t)$ and $D_{ij\cdot}(t)$ are not the same (and not equal). Also note that σ_D^2 , μ_D are <u>not</u> moments of differences of relative errors.

§3.12.2 Definitions and Data Sources

Adjusted taxes for a unit of local government are the total general purpose taxes imposed by the unit excluding taxes for schools and other educational purposes. The following specification holds.

t	EP	$D_{ij}(t)$, $D_{ijk}(t)$ refer to fiscal year
1,2	1,2	1970-1971
3	3	1970-1971
5	4	1971-1972
7	5	1972-1973
9	6	1973-1974
10	7	1974-1975

Table 3.8

Target Variables - Substate Adjusted Taxes

Except for EP 4, the data was obtained by the General Revenue Sharing surveys conducted by the Bureau of the Census. The surveys undertake complete enumeration of all local units of government. In EP 4 the data was obtained from the Census of Governments.

§3.12.3 Data Quality: Evidence

The major source of error is response variance, which arises partly from transferring and translating data from one format to another, e.g. "from...the form developed to be compatible with the particular report format for the State Controller's Office to the form required for the General Revenue Sharing Survey" (SRI 1974a, p.III-63).

A study of the accuracy of the adjusted tax estimates was performed by the Comptroller General (General Accounting Office 1975, p.25).

> To test the accuracy of the adjusted tax data we compiled adjusted taxes from local governments' financial records and compared our figures with those used...to calculate revenue sharing allocations for the fourth and fifth entitlement periods... Local officials often reported incorrect data to the Census Bureau because they misunderstood which revenue items should be included in compiling adjusted taxes. Other errors occurred because local officials did not submit data from the proper fiscal year.

A total of 111 local governments were sampled in each of EP's 4 and 5. It turned out that the smaller governments, i.e. those of population $\leq$ 2500, had the poorest data. In New Jersey the total dollar allocation to such governments (population $\leq$ 2500) affected by error in adjusted tax data was small. The main reasons for this are the small number of such governments and the small entitlements of such governments, so that even a large percentage error corresponds to a tiny absolute error. Accordingly we neglect the small governments sampled and focus attention on the remaining 44 local governments (88 observations). Assuming the Comptroller General estimates are error-free, the 88 observations from the two periods were distributed as:

11 no error

58 absolute relative error between 0.0% and 5.00%

19 absolute relative error greater than 5.00%.

No further information is available for the 58 "small" errors. Expressed as percents, the 19 "large" errors were 11.19, 40.03, 48.49, 18.70, 9.77, 11.69, 7.69, 26.10, 5.40, -8.35, 5.68, 14.87, -8.51, -14.65, -5.35, -15.14, 8.63, -5.99, -19.45. The sum of these 19 numbers is 130.8 and the sum of the squares is 6786.85.

§3.12.4 Data Quality: Assumptions and Conclusions

Assume that the 58 "small" relative errors are uniformly distributed on (-0.05, + 0.05). The following calculation shows that the expected sum of squares for the 58 "small" relative errors is 483.14×10^{-4}. Since each of the errors is uniform on (-0.05, + 0.05) its expected square is

$$\int_{.05}^{.05} x^2/0.1\, dx = 8.33 \times 10^{-4}$$

and $$(58)(8.33 \times 10^{-4}) = 483.14 \times 10^{-4} \quad .$$

Now consider the null hypothesis that the mean relative error is zero. The sample variance of the relative errors is

$$87^{-1}\ \{6786.85 + 483.14 - (130.8)^2/88\} \times 10^{-4}$$
$$= 81.328 \times 10^{-4} \quad .$$

Using a t-test at the 10% significance level we fail to reject the null hypothesis. Accordingly we use

$$\tilde{\mu}_D(i,j;t) = \tilde{\mu}_D(i,j,k;t) = 0 \quad .$$

§3.12.5 Relevant Distribution and Approximations

The discussion of §3.12.4 makes the following specification reasonable when the population exceeds 2500:

$$\tilde{\mu}_D(i,j;t) = \tilde{\mu}_D(i,j,k;t) = 0.0$$
$$\tilde{\sigma}_D^2(i,j;t) = \tilde{\sigma}_D^2(i,j,k;t) = (.09)^2 \quad .$$

§3.13 Intergovernmental Transfers (Substate level)

§3.13.1 Notation

Define

True (estimated) intergovernmental transfers of county government j in state i at time t $\quad G_{ij}(t)\ (\hat{G}_{ij}(t))$

True (estimated) intergovernmental transfers of subcounty government k in county j in state i at time t $\quad G_{ijk}(t)\ (\hat{G}_{ijk}(t))$.

Errors in GRS estimates of intergovernmental transfers are denoted by $g_{ij}(t) = \hat{G}_{ij}(t) - G_{ij}(t)$ and $g_{ijk}(t) = \hat{G}_{ijk}(t) - G_{ijk}(t)$ with estimates $\tilde{g}_{ij}(t)$, $\tilde{g}_{ijk}(t)$. The means and variances of $g_{ij}(t)/G_{ij}(t)$ and $g_{ijk}(t)/G_{ijk}(t)$ are denoted by $\mu_G(i,j;t)$, $\sigma_G^2(i,j;t)$, $\mu_G(i,j,k;t)$ and $\sigma_G^2(i,j,k;t)$, with estimates $\tilde{\mu}_G(i,j;t)$, $\tilde{\sigma}_G^2(i,j;t)$, $\tilde{\mu}_G(i,j,k;t)$ and $\tilde{\sigma}_G^2(i,j,k;t)$.

Note that $G_{ij}(t)$ is not the same as $G_{ij\cdot}(t)$. Also note that μ_G , σ_G^2 are <u>not</u> moments of differences of relative erros.

§3.13.2 Definition and Data Source

Intergovernmental transfers of revenue (Office of Revenue Sharing 1976, p.x) are

> ...amounts received by a unit of government from other governments...for use either for specific functions or for general financial support...
> The figure includes grants, shared taxes, contingent loans, and reimbursements for tuition costs, hospital care, construction costs, etc. Intergovernmental revenue does not include amounts received from sale of property or commodities, utilities services to other governments, or Federal general revenue sharing entitlement funds.

The following specification holds.

t	EP	$G_{ij}(t)$, $G_{ijk}(t)$ refer to fiscal year
1,2	1,2	1970-1971
3	3	1970-1971
5	4	1971-1972
7	5	1972-1973
9	6	1973-1974
10	7	1974-1975

Table 3.9

Target Variables - Intergovernmental Transfers

The intergovernmental transfer data was obtained by the same surveys as the adjusted taxes (see §3.12.2).

§3.13.3 Data Quality: Evidence

Intergovernmental transfers are collected by the same process as adjusted taxes and are subjected to the same kinds of edits as adjusted taxes (General Accounting Office 1975, p.26). The errors in the intergovernmental transfer data arise in the same way as errors in adjusted taxes (see §3.12.3).

§3.13.4 Data Quality: Assumptions and Conclusions

It is assumed that the distributions of relative errors for intergovernmental transfers are the same as those of adjusted taxes (see §3.13.4) because of the similar natures of the data collection procedures. However, no direct evidence is available.

§3.13.5 Relevant Distributions and Approximations

It follows that (see §3.12.5) when the population is greater than 2500 it is reasonable to use

$$\tilde{\mu}_G(i,j;t) = \tilde{\mu}_G(i,j;t) = 0$$

$$\tilde{\sigma}^2_G(i,j;t) = \tilde{\sigma}^2_G(i,j,k;t) = (.09)^2 \quad .$$

§3.14 Interrelationships Among State-level Data Elements

This section introduces and defines data elements that are functions of state-level data elements discussed in earlier sections. To keep notation simple, the argument t (denoting time) is suppressed. It is implicit that the same value of t applies uniformly to all data elements of which another data element may be composed. For example, $E_i(t) = T_i(t)/R_i(t)$ will be written $E_i = T_i/R_i$.

Relations between state-level data elements are also discussed.

§3.14.1 Notation

Define

True (estimated) <u>tax effort</u> of state i $\quad E_i = T_i/R_i \ (\hat{E}_i = \hat{T}_i/\hat{R}_i)$.

Various weighted averages of relative errors are important for using the delta method to analyze errors in state allocations (chapter 4). For vectors $\underset{\sim}{V},\underset{\sim}{W}$ such that $W_i \geq 0$ and $\Sigma W_i > 0$ let

$$\langle V;W\rangle = (\sum_i V_i W_i)/ \sum_i W_i \quad .$$

Now define the weighted average:

(3.45) $\langle p/P\rangle' = \langle p/P \,;\, PE/C\rangle$

(3.46) $\langle c/C\rangle' = \langle c/C \,;\, PE/C\rangle$

(3.47) $\langle t/T\rangle' = \langle t/T \,;\, PE/C\rangle$

(3.48) $\langle r/R\rangle' = \langle r/R \,;\, PE/C\rangle$

(3.49) $\langle p/P\rangle'' = \langle p/P \,;\, P/C\rangle$

(3.50) $\langle c/C\rangle'' = \langle c/C \,;\, P/C\rangle$

(3.51) $\langle t/T\rangle''' = \langle t/T \,;\, ET\rangle$

(3.52) $\langle r/R\rangle''' = \langle r/R \,;\, ET\rangle$.

Thus, for example, $\langle p/P\rangle' = \dfrac{\sum_i (P_iE_i/C_i)(p_i/P_i)}{\sum_i (P_iE_i/C_i)}$.

§3.14.3 Data Quality: Evidence

The weights used to derive the various weighted averages defined in §3.14.1 are highly collinear with the weights used to derive national averages. This collinearty can be estimated by the cosine of the angle between the vectors of weights computed from EP 1 data (Office of Revenue Sharing 1974a). The cosine of the angle between vectors $\underset{\sim}{V}$ and $\underset{\sim}{W}$ equals $\Sigma V_iW_i/((\Sigma V_i^2)(\Sigma W_i^2))^{1/2}$. The cosine of the angle between the vector of weights $P_i/P.$ and the vector of weights $(P_iE_i/C_i)/(PE/C).$ used to calculate $\langle p/P\rangle'$ was estimated to be .99. Similarly, the cosines of the angles between the vectors of

$[P_iE_i/C_i]/[(PE/C).]$ and $T_i/T.$, $R_i/R.$, $M_i/M.$ were .97, .98, .98;

$P_i/P.$ and $M_i/M.$, $(P_i/C_i)/(P/C).$ were .99, .99;

$E_iT_i/(ET).$ and $T_i/T.$, $R_i/R.$ were .99, .96;

$M_i/M.$ and $(P_i/C_i)/(P/C).$ was .96 .

§3.14.4 Data Quality: Assumptions

The near collinearity of the weights suggests that

$$\langle p/P\rangle' \quad , \quad \langle p/P\rangle'' \quad , \quad \text{and} \quad p./P.$$

are close with high probability, as are

$$\langle r/R\rangle' \quad , \quad \langle r/R\rangle''' \quad , \quad \text{and} \quad r./R. \quad ;$$
$$\langle t/T\rangle' \quad , \quad \langle t/T\rangle''' \quad , \quad \text{and} \quad t./T. \quad ; \quad \text{and}$$
$$\langle c/C\rangle' \quad , \quad \langle c/C\rangle'' \quad , \quad c./C. \quad , \quad \text{and} \quad c/C \quad .$$

Consider for example $p./P.$ and $\langle p/P\rangle' = \langle p/P;PE/C\rangle$.

Proposition 3.5 EP 1 data indicates that for any EP

$$E(\ \langle p/P\rangle' - p./P.)^2 \le 9.7 \times 10^{-8} \quad .$$

Proof: Let $X_i = P_i/P.$, $Y_i = (P_iE_i/C_i)/(PE/C)$ and notice

$$\begin{aligned} E(\ \langle p/P\rangle' - p./P.)^2 &= E(\Sigma\ Y_ip_i/P_i - \Sigma\ X_ip_i/P_i)^2 \\ &= E(\Sigma\ (p_i/P_i)(Y_i - X_i))^2 \\ &= E(\Sigma\ (p_i/P_i - E\ p_i/P_i)(Y_i - X_i) \\ &\quad + \Sigma\ (E\ p_i/P_i)(Y_i - X_i))^2 \end{aligned}$$

(because correlation between p_i/P_i and $p_j/P_j \le 0$ for $i \neq j$; see §3.2.4 and §3.2.5)

$$\begin{aligned} &\le \Sigma\ \mathrm{Var}(p_i/P_i)(Y_i - X_i)^2 \\ &\quad + (\Sigma\ (-A_i)(Y_i - X_i))^2 \end{aligned}$$

where A_i is the population undercoverage rate of state i . Calculations on EP 1 data show $\Sigma(Y_i - X_i)^2 = 5.7 \times 10^{-4}$ and

$$[\ \Sigma - A_i(Y_i - X_i)]^2 = 1.7 \times 10^{-8} \quad .$$

From §3.2.5 we have that for <u>any</u> EP

$$\mathrm{Var}(p_i/P_i) \leq 1.4 \times 10^{-4} \quad ,$$

consequently

$$E(\langle p/P\rangle' - p./P.)^2 \leq [(1.4)(5.7) + 1.7] \times 10^{-8} = 9.7 \times 10^{-8}$$

*****

Thus $\langle p/P\rangle'$ and $p./P.$ are close with high probability. The force of this closeness will be discussed in chapter 4, section 4.

Now consider covariances between data elements. It is usually the case that stochastic variability in different data elements arises from distinct sources. For example, random variability of state population estimates arises primarily from errors in estimates of net interstate migration while variability in estimates of net state and local taxes arises primarily from sampling variance. Bear in mind that population undercounts are fixed effects and irrelevant to covariance considerations. We shall therefore assume that the covariances between

(3.53a) $\quad p_i/P_i - \langle p/P\rangle' \quad , \quad t_i/T_i - \langle t/T\rangle' \quad , \quad r_i/R_i - \langle r/R\rangle' \quad ,$

and $\quad c_i/C_i - \langle c/C\rangle'$

are negligible, as are the covariances between

(3.53b) $\quad p_i/P_i - \langle p/P\rangle'' \quad , \quad t_i/T_i - \langle t/T\rangle''' \quad , \quad r_i/R_i - \langle r/R\rangle''' ,$

$c_i/C_i - \langle c/C\rangle''$ and $u_i/U_i - u./U. \quad .$

The only questionable part of the above assumption concerns the covariance between population and per capita income. The randomness arises solely from the respective updating procedures, and we restrict our attention to possible correlations between the respective updates. We note (c.f. (3.20)) that $\mathrm{Cov}(p_i/P_i \,,\, c_i/C_i)$ is approximately equal to $\mathrm{Cov}(p_i/P_i \,,\, m_i/M_i) - \mathrm{Var}(p_i/P_i)$. Evidence about the magnitude of $\mathrm{Cov}(p_i/P_i \,,\, m_i/M_i)$ is completely lacking. We do not even know whether the covariance is positive or negative. The assumption that $\mathrm{Cov}(p_i/P_i \,,\, c_i/C_i)$ is negligible is thus one of ignorance.

Finally note that although both $\hat{I}_i$ and $\hat{T}_i$ contain error from $\hat{K}_i$, the periods of reference for I_i, T_i do not coincide and therefore $\hat{I}_i$, $\hat{T}_i$ are uncorrelated.

§3.15 Interrelationships Among Substate Data Elements

This section introduces and defines data elements that are functions of substate-level data elements discussed in earlier sections. To keep notation simple, the argument t (denoting time) is supressed (as in §3.14). Interrelations between substate-level data elements are also discussed.

§3.15.1 Notation

Define

True (estimated) tax effort of county area j in state i

$$(D_{ij} + D_{ij}.)/(P_{ij}C_{ij}) \quad (\hat{D}_{ij} + \hat{D}_{ij}.)/(\hat{P}_{ij}\hat{C}_{ij}))$$

True (estimated) tax effort of local unit k in county area j in state i

$$(D_{ijk}/(P_{ijk}C_{ijk}) \quad (\hat{D}_{ijk}/(\hat{P}_{ijk}\hat{C}_{ijk}))$$

True (estimated) relative income of county area j in state i

$$C_i/C_{ij} \quad (\hat{C}_i/\hat{C}_{ij})$$

True (estimated) relative income of local unit k in county area j in state i

$$C_{ij}/C_{ijk} \quad (\hat{C}_{ij}/\hat{C}_{ijk}) \ .$$

For New Jersey the set K of local area subscripts is partitioned into townships $K1$ and places and municipalities $K2$. For $\alpha = 1,2$ elements of $K\alpha$ will be denoted by $k\alpha$ or simply by k. Recall that J denotes the set of county subscripts for New Jersey.

Various weighted averages of relative errors are important for using the delta method to analyze errors in substate allocations. We shall suppress the subscript i because all substate analysis refers to New Jersey. Defining

(3.54) $$w_j = [(D_j + D_j.)/ C_j^2]/ \sum_J [(D_j + D_j.)/C_j^2]$$

and

(3.55) $$w_{jk\alpha} = [D_{jk\alpha}/C^2_{jk\alpha}]/ \sum_{\ell\epsilon K\alpha} [D_{j1\alpha}/C^2_{jk\alpha}]$$

we let

(3.56) $$<d/D>^* = \sum_J w_j(d_j + d_j.)/(D_j + D_j.)$$

(3.57) $$<d/D>^{**}_{j\alpha} = \sum_{K\alpha} w_{jk\alpha} d_{jk\alpha}/D_{jk\alpha}$$

(3.58) $$<c/C>^* = \sum_J w_j(c_j/C_j)$$

(3.59) $$<c/C>^{**}_{j\alpha} = \sum_{K\alpha} w_{jk\alpha} c_{jk\alpha}/C_{jk\alpha} \quad .$$

The following notation will be utilized in chapter 5.

(3.60) $$\xi_j = (d_j + d_j.)/(D_j + D_j.) - <d/D>^* -2(c_j/C_j - <c/C>^*),$$

and

(3.61) $$\xi_{jk\alpha} = (d_{jk\alpha}/D_{jk\alpha}) - <d/D>^{**}_{j\alpha} - 2[c_{jk\alpha}/C_{jk\alpha} - <c/C>^{**}_{j\alpha}] ,$$

and

(3.62) $$\phi_{j\alpha} = (\sum_{K\alpha} d_{jk\alpha})/(\sum_{K\alpha} D_{jk\alpha}) - (d_j + d_j.)/(D_j + D_j.) \quad .$$

In this section we shall write c/C to represent

$c_i/C_i = \dfrac{\hat{C}_i - C_i}{C_i}$ for i corresponding to New Jersey. Elsewhere c/C refers to the relative error for national per capita income.

§3.15.2 and §3.15.3 Definitions and Data Sources; Data Quality: Evidence

There are no new definitions or data sources to consider since all quantities defined in §3.15.1 are composed entirely of data elements discussed in previous sections.

§3.15.4 Data Quality: Assumptions

Recall that §3.13.4 discussed how the near collinearity of the weights involved in state-level weighted averages $\langle p/P\rangle^*$ and $p./P.$ implied that these weighted averages were close with high probability. Because collinearity is also present between the weights implicit in the county-level weighted averages $\langle c/C\rangle^*$ and c/C, these two weighted averages will also be close with high probability. This closeness with be exploited in §5.4.

Because errors in adjusted taxes and intergovernmental transfers arise from response variability on the part of local government officials, it is assumed that for any two units of local government (possibly the same)

(i) the estimate of adjusted taxes of one unit is uncorrelated with estimates of intergovernmental transfers, population, and per capita income of the other

and

(ii) the estimate of intergovernmental transfers of one unit is uncorrelated with estimates of population and per capita income of the other.

One might anticipate that the absolute errors in $\hat{G}_{ij}$ and $\hat{D}_{ij}$ (or $\hat{G}_{ijk}$ and $\hat{D}_{ijk}$) would be correlated, since if local officials misunderstood the instructions for calculating adjusted taxes they would also be confused about intergovernmental transfers. However, and this is what matters, for our purposes, the <u>actual</u> errors will not be correlated. For different units of local government, adjusted taxes are uncorrelated and intergovernmental transfers are uncorrelated. As in §3.13.4 we assume that the updating errors in population and per capita income are uncorrelated; in particular we assume

(3.63) $\quad \mathrm{Cov}(p_j/P_j - p./P.\ ,\ c_{jk\alpha}/C_{jk\alpha} - \langle c/C\rangle^{**}_{j\alpha}) = 0$.

To summarize, we have the following situation: for counties j , j' in New Jersey (possibly the same) and for local areas k , k' in New Jersey (possibly the same) each of

$$d_j\ ,\ d_{jk}\ ,\ g_j\ ,\ g_{jk}\ ,\ c_{jk\alpha}/C_{jk\alpha} - \langle c/C\rangle^{**}_{j\alpha}\ ,$$

is uncorrelated with each of

$$d_{j'}\ ,\ d_{j'k'}\ ,\ g_{j'}\ ,\ g_{j'k'}\ ,\ c_{j'}/C_{j'} - \langle c/C\rangle^{*}\ ,\ p_{j'}/P_{j'} - p./P.$$

except when $j = j'$ then $\mathrm{Cov}(d_j/D_j\ ,\ d_{j'}/D_{j'}) = \sigma_D^2(j)$ and $\mathrm{Cov}(g_j/G_j\ ,\ g_{j'}/G_{j'}) = \sigma_G^2(j)$ and when $jk = j'k'$ then $\mathrm{Cov}(d_{jk}/D_{jk}\ ,\ d_{j'k'}/D_{j'k'}) = \sigma_D^2(j,k)$ and $\mathrm{Cov}(g_{jk}/G_{jk}\ ,\ g_{j'k'}/G_{j'k'}) = \sigma_G^2(j,k)$.

§3.16 Outline - Symbol - Glossary for Chapter 3

The following outline summarizes the notation of chapter 3 and provides references to the sections introducing the notation. Subscripts and "t" for time are omitted below unless required for clarity.

Subscripts (§3.1.1)

i	state
j	county
k	subcounty area
k1	township
k2	municipality or place
b	black
w	white
J	collection of county subscripts for New Jersey
K	collection of subcounty subscripts for New Jersey
$K\alpha$	for $\alpha = 1$ all townships in K ; for $\alpha = 2$ all places and municipalities in K

Conventions (§3.1 and §3.1.3-4)

^	estimate used in allocation
~	estimate used to evaluate errors in allocation
Z	upper case letters denote parameters
$z=\hat{Z}-Z$	lower case letters denote the difference between the estimate and the parameter

$$\Delta_Z(i;t) = z_i(t) - z_i(1)$$

$$= \hat{Z}_i(t) - Z_i(t) - (\hat{Z}_i(1) - Z_i(1))$$

$$= \hat{Z}_i(t) - \hat{Z}_i(1) - (Z_i(t) - Z_i(1))$$

$\cdot$ = sum on subscript, e.g. $Z. = \Sigma Z_i$

$\langle V;W\rangle = \Sigma V_i(W_i/W.)$

<u>Variables</u> (the numbers in parentheses refer to the sections in the text introducing the notation)

P	population (§3.2.1, §3.4.1)
A	relative population undercount (= $-E_P/P$) (§3.2.1, §3.3.1)
U	urbanized population (§3.3.1)
M	money income (§3.5.1, §3.6.1)
C	per capita income (§3.5.1, §3.6.1)
R	personal income (§3.7.1)
T	net state and local taxes (§3.8.1)
K	state individual income tax collection (§3.9.1)
L	federal individual income tax liabilities (§3.10.1)
I	income tax amount (=median {.01L, .15K, .06L}) (§3.11.1)
D	adjusted taxes (§3.12.1)
G	intergovernmental transfers (§3.13.1)
E	state-level tax effort (=T/R) (§3.14.1)
$\dfrac{D_{ij}+D_{ij\cdot}}{P_{ij}C_{ij}}$	county level tax effort (§3.15.1)
$D_{ijk}/(P_{ijk}C_{ijk})$	subcounty-level tax effort (§3.15.1)
$\dfrac{C_i}{C_{ij}}$, $\dfrac{C_{ij}}{C_{ijk}}$	relative per capita income (§3.15.1)

<u>More Conventions</u> (α takes on values 1 and 2)

$\langle Z\rangle' = \langle Z;PE/C\rangle$ (§3.14.1)

$\langle Z\rangle'' = \langle Z;P/C\rangle$ (§3.14.1)

$\langle Z\rangle''' = \langle Z;ET\rangle$ (§3.14.1)

The remaining conventions are defined in §3.15.1.

$w_j \quad = [(D_j+D_j.)/C_j^2]/\sum_{\ell\varepsilon J}[(D_\ell+D_\ell.)/C_\ell^2]$

$w_{jk\alpha} \quad = [D_{jk\alpha}/C_{jk\alpha}^2]/\sum_{\ell\varepsilon K\alpha}[D_{j\alpha}/C_{j\ell\alpha}^2]$

$\langle d/D\rangle^* \quad = \sum_J w_j(d_j+d_j.)/(D_j+D_j.)$

$\langle d/D\rangle_{j\alpha}^{**} \quad = \sum_{K\alpha} w_{jk}(d_{jk\alpha}/D_{jk\alpha})$

$\langle c/C\rangle^* \quad = \sum_J w_j(c_j/C_j)$

$\langle c/C\rangle_{j\alpha}^{**} \quad = \sum_{K\alpha} w_{jk\alpha}(c_{jk\alpha}/C_{jk\alpha})$

$\xi_j \quad = (d_j+d_j.)/(D_j+D_j.) - \langle d/D\rangle^* - 2(c_j/C_j - \langle c/C\rangle^*)$

$\xi_{jk\alpha} \quad = (d_{jk\alpha}/D_{jk\alpha}) - \langle d/D\rangle_{j\alpha}^{**} - 2(c_{jk\alpha}/C_{jk\alpha} - \langle c/C\rangle_{j\alpha}^{**})$

$\phi_{j\alpha} \quad = (\sum_{K\alpha} d_{jk})/(\sum_{K\alpha} D_{jk\alpha}) - (d_j+d_j.)/(D_j+D_j.)$

<u>Moments for state-level data elements</u>

$\mu_Z(i)$ mean of functions of relative errors

$\sigma_Z^2(i)$ variance of function of relative errors

σ_{ZZ} Covariance between functions of relative errors

where the function is $\frac{z_i}{Z_i} - \frac{z.}{Z.}$ for $Z_i = P_i$, U_i, C_i, T_i, R_i, and I_i; and the function is $\frac{z_i}{Z_i}$ for $Z_i = K_i$ and L_i.

<u>Moments for substate-level data elements</u>

$\mu_P(i,j), \sigma_P^2(i,j)$	moments of $p_{ij}/P_{ij} - p_{i\cdot}/P_{i\cdot}$
$\mu_P(i,j,k), \sigma_P^2(i,j,k)$	moments of $p_{ijk}/P_{ijk} - p_{i\cdot\cdot}/P_{i\cdot\cdot}$
$\mu_C(i,j), \sigma_C^2(i,j)$	moments of $c_{ij}/C_{ij} - c_i/C_i$
$\mu_C(i,j,k), \sigma_C^2(i,j,k)$	moments of c_{ijk}/C_{ijk}
$\mu_D(i,j), \sigma_D^2(i,j)$	moments of d_{ij}/D_{ij}
$\mu_D(i,j,k), \sigma_D^2(i,j,k)$	moments of d_{ijk}/D_{ijk}
$\mu_G(i,j), \sigma_G^2(i,j)$	moments of g_{ij}/G_{ij}
$\mu_G(i,j,k), \sigma_G^2(i,j,k)$	moments of g_{ijk}/G_{ijk}

Chapter 4 Interstate Allocations in GRS

§4.1 Introduction

The Legislation (Public Law 92-512) precisely specifies the procedure to compute GRS allocations to states. Although the specification relies on words rather than algebraic formulas it is unambiguous. Considerable debate and discussion preceded the adoption of the procedure, which reflects a compromise between the allocation formulas initially proposed by the House and by the Senate.* The procedure sets the GRS allocation to state i proportional to the larger of the "House" amount H_i and the "Senate" amount S_i given by

$$(4.1)\qquad H_i = \frac{35}{159}\left(\frac{P_i}{P.} + \frac{P_i/C_i}{(P/C).} + \frac{U_i}{U.}\right) + \frac{27}{159}\left(\frac{I_i}{I.} + \frac{E_iT_i}{(ET).}\right)$$

and

$$(4.2)\qquad S_i = \frac{P_iE_i/C_i}{(PE/C).}$$

where

P_i = population

U_i = urbanized population

C_i = per capita income

I_i = median $\{.01L_i\ ,\ .15K_i\ ,\ .06L_i\}$ = income tax amount

---

* Dommel (1974) gives a lively account of the political and legislative history of revenue sharing. A broad comprehensive description of the background and distributional effects of the GRS program is found in Nathan et al. (1974).

L_i = federal individual income tax liabilities

K_i = state individual income tax collections

$E_i = T_i/R_i$ = tax effort

T_i = net state and local tax collections

and R_i = personal income

are state-level data elements discussed in chapter 3. The presentation of the formulas for H_i and S_i by the legislation is algebraically equivalent to (4.1), (4.2) above.* The portion of the pie allocated to State i is** $Y_i = X_i/X.$, where $X_i = \max(S_i , H_i)$. The size of the pie was essentially fixed under the original GRS program.

---

* The complexity of the House formula (4.1) and the substate allocation formulas (see chapter 5) arises from the considerable political difficulties the House Ways and Means Committee had in producing an acceptable formula. Representative James C. Corman, a member of the Ways and Means Committee who had opposed the bill, described the emergence of the House formula in this way (Congressional Record, June 22, 1972 (Daily Edition), p.H5949): "We finally quit, not because we hit on a rational formula, but because we were exhausted. And finally we got one that almost none of us could understand at the moment. We were told that the statistics were not available to run the [computer] print on it. So we adopted it, and it is here for you today."

** For Alaska and Hawaii the procedure was somewhat different. To account for generally higher price levels, "non-contiguous state adjustment factors," say F_{Alaska} , F_{Hawaii} were defined. For entitlement periods 1-6 they were 1.25, and 1.15 respectively. For i denoting Alaska and Hawaii, If $S_iF_i > H_i$ then X_i is set equal to S_i and

to the final allocation an extra 25% or 15% is taken from a source of funds earmarked "non-contiguous states adjustment amounts". If the source does not contain enough money, the 25% and 15% adjustments are scaled down. These adjustments need not be considered until chapter 6.

Large biases in the allocations arise from biases in population estimates $\hat{P}_i$, per capita income estimates $\hat{C}_i$, and income tax amount estimates $\hat{I}_i$. Note that if the uncounted population in a state had the same per capita income as the rest of the state then population undercoverage would not bias the estimate of per capita income. In fact, the uncounted populations tend to have lower than average per capita income and so population undercoverage causes upward biases in per capita income estimates. The effect of population biases is thus magnified by the presence of C_i^{-1} terms in (4.1), (4.2).

The effort in this chapter is aimed at providing approximations for $\hat{Y}_i - Y_i$, $E(\hat{Y}_i - Y_i)$, $Var(\hat{Y}_i)$, and $E|\hat{Y}_i - Y_i|$. These approximations will be used in chapter 6 to examine expected losses from perturbations in GRS allocations that arise from errors in data. Extensive notation is required (§4.2) before these approximations can be presented (§4.3). Assumptions and magnitudes of neglected terms are discussed later (§4.4), and then the derivations are given (§4.5).

§4.2 Notation

It is convenient to partition the states into those that should be using the House formula and those that should be using the Senate formula. Let

$$
(4.3) \qquad
\begin{aligned}
X_i &= \max(S_i \, , \, H_i) & \hat{X}_i &= \max(\hat{S}_i \, , \, \hat{H}_i) \\
Y_i &= X_i / X. & \hat{Y}_i &= \hat{X}_i / \hat{X}. \\
\mathcal{S} &= \{i\colon X_i = S_i\} & \hat{\mathcal{S}} &= \{i\colon \hat{X}_i = \hat{S}_i\} \\
\mathcal{H} &= \{i\colon X_i = H_i\} \, . & \hat{\mathcal{H}} &= \{i\colon \hat{X}_i = \hat{H}_i\} \, .
\end{aligned}
$$

The argument t denoting time will be suppressed in this chapter because each quantity is to be evaluated for the same entitlement period. Much reference will be made to the various state-level data elements discussed in chapter 3. The following table will assist the reader in checking the definitions and evaluating the models $(\mu_P(i) \, , \, \sigma_P^2(i) \, , \, \sigma_{PP}(i,j) \, , \, \mu_C(i) \, , \ldots)$ of the various data elements $(P_i \, , \, C_i \, , \text{ etc.})$

Notation	Name of Data Element	Location of Definition	Location of Error Model
P_i	population	§3.2.2	§3.2.5
U_i	urbanized population	§3.3.2	§3.3.5
C_i	per capita income	§3.5.2	§3.5.5
R_i	personal income	§3.7.2	§3.7.5
T_i	net state and local taxes	§3.8.2	§3.8.5
I_i	income tax amount	§3.11.2	§3.11.5

Table 4.1

Locator for Description of Data

Expressions for covariances between state-level data elements are simplified by the following notation. For a generic data element or function of data elements Z and for i,j,k taking values $1,2,\ldots,n$ let

$$(4.4) \qquad \sigma^*(Z,i,j) = -(Z_i/Z.)\,\mathrm{Var}(\hat{Z}_i/Z_i) - (Z_j/Z.)\,\mathrm{Var}(\hat{Z}_j/Z_j) + \sum_k (Z_k/Z.)^2\,\mathrm{Var}(\hat{Z}_k/Z_k) \quad .$$

Using corollary 2.2 we may interpret $\sigma^*(Z,i,j)$ as an approximation to the relative covariance between $\hat{Z}_i/\hat{Z}.$ and $\hat{Z}_j/\hat{Z}.$ for $i \neq j$, i.e. the covariance between $(Z_i/Z.)^{-1}\,(\hat{Z}_i/\hat{Z}.)$ and $(Z_j/Z.)^{-1}\,(\hat{Z}_j/\hat{Z}.)$.

The succeeding notation will clarify the presentation of biases and variances of errors in allocation. Refer to Table 4.1 to locate necessary definitions and evaluations. We define

$$(4.5) \qquad \mu_Y(i) = E\left(\frac{\hat{Y}_i - Y_i}{Y_i}\right)$$

$$(4.6) \qquad \sigma_Y^2(i) = \mathrm{Var}(\hat{Y}_i/Y_i)$$

$$(4.7) \qquad \sigma_{YY}(i,j) = \mathrm{Cov}(\hat{Y}_i/Y_i \, , \, \hat{Y}_j/Y_j) \quad .$$

The bias in $(\hat{S}_i - S_i)/S_i$ wll be approximated by

$$(4.8) \qquad \mu_S(i) = \mu_P(i) - \mu_C(i)$$

where we recall $\mu_P(i)$ [or $\mu_C(i)$] is an approximation for the bias in $(\hat{P}_i - P_i)/P_i$ [or $(\hat{C}_i - C_i)/C_i$] . The variance of $\hat{S}_i/S_i$ will be approximated by

$$\sigma_S^2(i) = \sigma_P^2(i) + \sigma_T^2(i) + \sigma_C^2(i) + \sigma_R^2(i) \tag{4.9}$$

where the terms on the right are approximations for the variances of

$$\frac{\hat{P}_i - P_i}{P_i}, \quad \frac{\hat{T}_i - T_i}{T_i}, \quad \frac{\hat{C}_i - C_i}{C_i}, \quad \text{and} \quad \frac{\hat{R}_i - R_i}{R_i} \quad .$$

The bias and variance of $(\hat{H}_i - H_i)/H_i$ will be approximated by

$$\mu_H(i) = H_i^{-1} \left\{ \frac{35}{159} \left\{ \left(\frac{P_i}{P.} + \frac{P_i/C_i}{(P/C).}\right) \mu_P(i) + \frac{U_i}{U.} \mu_U(i) - \frac{P_i/C_i}{(P/C).} \mu_C(i) \right\} + \frac{27}{159} \frac{I_i}{I.} \mu_I(i) \right\} \tag{4.10}$$

and

$$\sigma_H^2(i) = H_i^{-2} \left\{ \left(\frac{35}{159}\right)^2 \left\{ \left(\frac{P_i}{P.} + \frac{P_i/C_i}{(P/C).}\right)^2 \sigma_P^2(i) + \left(\frac{U_i}{U.}\right)^2 \sigma_U^2(i) + \left(\frac{P_i/C_i}{(P/C).}\right)^2 \sigma_C^2(i) \right\} + \left(\frac{27}{159}\right)^2 \left\{ \frac{T_i^2/R_i^2}{(T^2/R).} \left(4\, \sigma_T^2(i) + \sigma_R^2(i)\right) + \left(\frac{I_i}{I.}\right)^2 \sigma_I^2(i) \right\} \right\} \tag{4.11}$$

where $\mu_U(i)$ and $\mu_I(i)$ $[\sigma_U^2(i)$ and $\sigma_I^2(i)]$ are approximations for the biases [variances] of $(\hat{U}_i - U_i)/U_i$ and $(\hat{I}_i - I_i)/I_i$.

Approximations for $\mathrm{Cov}(\hat{S}_i/S_i\ ,\ \hat{S}_j/S_j)$, $\mathrm{Cov}(\hat{S}_i/S_i\ ,\ \hat{H}_j/H_j)$ and $\mathrm{Cov}(\hat{H}_i/H_i\ ,\ \hat{H}_j/H_j)$ $i \neq j$ are given by $\sigma_{SS}(i,j)$, $\sigma_{SH}(i,j)$ and $\sigma_{HH}(i,j)$ defined by

(4.12) $$\sigma_{SS}(i,j) = \sigma^*(S,i,j)$$

where σ^* is defined by (4.4),

(4.13) $$\sigma_{SH}(i,j) = H_j^{-1} \left\{ \frac{35}{159} \frac{P_j}{P.} \sigma_{PP}(i,j) + \frac{35}{159} \frac{P_j/C_j}{(P/C).} (\sigma_{PP}(i,j) + \sigma_{CC}(i,j)) + \frac{27}{159} \frac{T_j^2/R_j}{(T^2/R).} (2\,\sigma_{TT}(i,j) + \sigma_{RR}(i,j)) \right\} ,$$

and

(4.14) $$\sigma_{HH}(i,j) = H_i^{-1} H_j^{-1} \left\{ \left(\frac{35}{159}\right)^2 \left[\left(\frac{P_j}{P.} + \frac{P_j/C_j}{(P/C).}\right)\left(\frac{P_i}{P.} + \frac{P_i/C_i}{(P/C).}\right) \sigma_{PP}(i,j) + \left(\frac{U_i}{U.}\right)\left(\frac{U_j}{U.}\right) \sigma_{UU}(i,j) + \left(\frac{P_i/C_i}{(P/C).}\right)\left(\frac{P_j/C_j}{(P/C).}\right) \sigma_{CC}(i,j) \right] + \left(\frac{27}{159}\right)^2 \left[\left(\frac{T_i^2/R_i}{(T^2/R).}\right)\left(\frac{T_j^2/R_j}{(T^2/R.}\right) (4\sigma_{TT}(i,j) + \sigma_{RR}(i,j)) + \frac{I_i}{I.} \frac{I_j}{I.} \sigma^*(I,i,j) \right] \right\} .$$

Recall that $\sigma_{PP}(i,j)$, $\sigma_{CC}(i,j)$, $\sigma_{TT}(i,j)$ etc. are approximations for $Cov(\frac{\hat{P}_i-P_i}{P_i} , \frac{\hat{P}_j-P_j}{P_j})$, $Cov(\frac{\hat{C}_i-C_i}{C_i} , \frac{\hat{C}_j-C_j}{C_j})$, $Cov(\frac{\hat{T}_i-T_i}{T_i} , \frac{\hat{T}_j-T_j}{T_j})$ etc.

§4.3 Conclusions

In §4.5 the following approximations will be derived (recall that $\langle p/P\rangle'$, $\langle t/T\rangle'$, etc. are defined by (3.45)-(3.52)):

$$\frac{\hat{S}_i-S_i}{S_i} = (p_i/P_i - \langle p/P\rangle') + (t_i/T_i - \langle t/T\rangle') - (c_i/C_i - \langle c/C\rangle') - (r_i/R_i - \langle r/R\rangle') + o_p(N^{-1/2}) \tag{4.15}$$

$$(4.16)\qquad \frac{\hat{H}_i - H_i}{H_i} = \frac{35}{159}\, H_i^{-1} \left\{ \frac{P_i}{P.}\left(\frac{p_i}{P_i} - \frac{p.}{P.}\right) + \frac{P_i/C_i}{(P/C).}\left(\frac{p_i}{P_i} - \langle p/P\rangle''\right) \right.$$

$$+ \frac{U_i}{U.}\left(\frac{u_i}{U_i} - \frac{u.}{U.}\right) - \left.\frac{P_i/C_i}{(P/C).}\left(\frac{c_i}{C_i} - \langle c/C\rangle''\right)\right\}$$

$$+ \frac{27}{159}\, H_i^{-1} \left\{ 2\,\frac{T_i^2/R_i}{(T^2/R).}\left(\frac{t_i}{T_i} - \langle t/T\rangle'''\right)\right.$$

$$+ \frac{I_i}{I.}\left(\frac{i_i}{I_i} - \frac{i.}{I.}\right)$$

$$- \left.\frac{T_i^2/R_i}{(T^2/R).}\left(\frac{r_i}{R_i} - \langle r/R\rangle'''\right)\right\} + o_p(N^{-1/2})$$

$$(4.17)\qquad \frac{\hat{Y}_i - Y_i}{Y_i} = -\frac{\sum_{\mathcal{S}}(\hat{S}_j - S_j) + \sum_{\mathcal{H}}(\hat{H}_j - H_j)}{\sum_{\mathcal{S}} S_j + \sum_{\mathcal{H}} H_j} + \begin{cases} \dfrac{\hat{S}_i - S_i}{S_i} & i \text{ in } \mathcal{S} \\[2ex] \dfrac{\hat{H}_i - H_i}{H_i} & i \text{ in } \mathcal{H} \end{cases} + o_p(N^{-1/2}).$$

Formula (4.17) is useful for examining the sensitivity of allocations to small perturbations in the data. The first term on the right hand side of (4.17) is smaller than the second so temporarily let ε denote its sum with the $o_p(N^{-1/2})$ term. In the remainder of this section ε will denote a variety of terms, all of which will for simplicity be neglected for now. Sometimes ε will be random and sometimes nonrandom. Discussion of the magnitude of ε appears in §4.4, but the analysis is heuristic since we do not demonstrate that the retained terms are a larger order of magnitude than the neglected terms.

The terms $\langle p/P\rangle'$, $\langle t/T\rangle'$, etc. are weighted averages of relative errors defined by (3.45)-(3.52). Since $\langle p/P\rangle'$ and $\langle t/T\rangle'$ are close to $p./P.$ and $t./T.$ and c_i/C_i is close to $m_i/M_i - p_i/P_i$ (by relation (3.11)) we use (4.15) to show that for states i in S (i.e. using the Senate formula)

$$(4.18) \qquad (\hat{Y}_i - Y_i)/Y_i = 2(p_i/P_i - p./P.) + (t_i/T_i - t./T.)$$

$$- (m_i/M_i - m./M.) - (r_i/R_i - r./R.) + \varepsilon \quad .$$

Formula (4.18) suggests that the effect of a relative error in population is approximately twice as severe as for another data element. Actually the effect is slightly less than twice as great because errors in population and money income may offset each other. For example, if the uncounted population had the same per capita income as the rest of the state, errors in m_i/M_i (aside from response error in $\hat{M}_i$) and p_i/P_i would cancel out. It is emphasized, however, that errors in population do affect per capita income estimates (because poor people tend to have higher undercoverage rates) and the conclusion of the Stanford Research Institute (1974b, p.IV-11), that each of the data inputs in (4.15) or (4.22) is equally important for states i in S is not warranted.

For states whose allocations were based on the House formula, (4.16) gives

$$\frac{\hat{Y}_i - Y_i}{Y_i} = \frac{35}{159Y_i} \left\{ \left(\frac{P_i}{P.} + 2\,\frac{P_i/C_i}{(P/C).}\right)\left(\frac{p_i}{P_i} - \frac{p.}{P.}\right) + \frac{U_i}{U.}\left(\frac{u_i}{U_i} - \frac{u.}{U.}\right) - \frac{P_i/C_i}{(P/C).}\left(\frac{m_i}{M_i} - \frac{m.}{M.}\right) \right\}$$

$$+ \frac{27}{159Y_i} \left\{ 2\,\frac{T_i^2/R_i}{(T^2/R).}\left(\frac{t_i}{T_i} - \frac{t.}{T.}\right) + \frac{I_i}{I.}\left(\frac{i_i}{I_i} - \frac{i.}{I.}\right) - \frac{T_i^2/R_i}{(T^2/R).}\left(\frac{r_i}{R_i} - \frac{r.}{R.}\right) \right\} + \varepsilon \quad .$$

Since errors in population estimates tend to affect urbanized population the same way, we have for i in H

(4.19)
$$\frac{\hat{Y}_i - Y_i}{Y_i} = \frac{35}{159Y_i} \left\{ \left(\frac{P_i}{P.} + 2\,\frac{P_i/C_i}{(P/C).} + \frac{U_i}{U.}\right)\left(\frac{p_i}{P_i} - \frac{p.}{P.}\right) - \frac{P_i/C_i}{(P/C).}\left(\frac{m_i}{M_i} - \frac{m.}{M.}\right) \right\}$$

$$+ \frac{27}{159Y_i} \left\{ 2\,\frac{T_i^2/R_i}{(T^2/R).}\left(\frac{t_i}{T_i} - \frac{t.}{T.}\right) + \frac{I_i}{I.}\left(\frac{i_i}{I_i} - \frac{i.}{I.}\right) - \frac{T_i^2/R_i}{(T^2/R).}\left(\frac{r_i}{R_i} - \frac{r.}{R.}\right) \right\} + \varepsilon \quad .$$

The effects of relative errors in data elements are harder to compare for the House formula than for the Senate formula because the coefficients $P_i/P.$, $(P_i/C_i)/(P/C).$, $U_i/U.$, $(T_i^2/R_i)/(T^2/R).$ and $I_i/I.$ do not vary over states in a fixed manner. However "on the average" each coefficient is the same size, since each lies between 0 and 1 and sums over states to 1. With this perspective, relative errors in population are the most critical because they are multiplied in (4.19) by four coefficients, each weighted by 35/159. Next in importance are relative errors in net state and local taxes $(\hat{T}_i)$, which are multiplied by two coefficients, each weighted by 27/159. Money income $(\hat{M}_i)$ follows, multiplied by one coefficient weighted by 35/159. Least in importance are income tax amount $(\hat{I}_i)$ and personal income $(\hat{R}_i)$ relative errors, since each is multiplied by one coefficient weighted by 27/159. Except for their interpretation of the effects of errors in population upon per capita income, the conclusions of the Stanford Research Institute (1974b, pp.IV 11, IV-12) about relative importance of data elements for states i in H are supported by this heuristic analysis.

Formulas (4.20), (4.21) and (4.23) give the relative biases, variances and mean absolute relative errors in state allocations:

$$(4.20)\qquad \mu_Y(i) = E\left(\frac{\hat{Y}_i - Y_i}{Y_i}\right) = -\frac{\sum_{S} S_j \mu_S(j) + \sum_{H} H_j \mu_H(j)}{\sum_{S} S_j + \sum_{H} H_j} + \begin{cases} \mu_S(i) & i \text{ in } S \\ \mu_H(i) & i \text{ in } H \end{cases} + o(N^{-1/2})$$

$$(4.21)\qquad \sigma_Y^2(i) = \mathrm{Var}(\hat{Y}_i/Y_i) = \mathrm{Var}(\hat{X}_i/X_i) + \sum_{j,k} Y_j Y_k \,\mathrm{Cov}(\hat{X}_j/X_j\, , \hat{X}_k/X_k) - 2\sum_j Y_j \,\mathrm{Cov}(\hat{X}_i/X_i\, , \hat{X}_j/X_j) + o(N^{-1})$$

where

$$\mathrm{Var}(\hat{X}_i/X_i) = \sigma_H^2(i) + o(N^{-1}) \qquad \text{for } i \text{ in } H$$

$$= \sigma_S^2(i) + o(N^{-1}) \qquad \text{for } i \text{ in } S$$

and

(4.22) $\mathrm{Cov}(\hat{X}_i/X_i\ ,\ \hat{X}_j/X_j) = \sigma_{HH}(i,j) + o(N^{-1})$ for i,j in H, $i \neq j$

$= \sigma_{HS}(i,j) + o(N^{-1})$ for i in H and j in S

$= \sigma_{SS}(i,j) + o(n^{-1})$ for i,j in S, $i \neq j$

$= \sigma_H^2(i) + o(N^{-1})$ if i=j in H

$= \sigma_S^2(i) + o(N^{-1})$ if i=j in S

and

(4.23) $$E\left|\frac{\hat{Y}_i - Y_i}{Y_i}\right| = 2\,\sigma_Y(i)\,\phi(-\mu_Y(i)/\sigma_Y(i))$$

$$+\ \mu_Y(i)(1-2\,\Phi(-\mu_Y(i)/\sigma_Y(i)) + o(N^{-1/2})\ ,$$

where ϕ and Φ denote the normal probability density and distribution functions.

For i in S we have

$$E(\hat{Y}_i - Y_i)/Y_i = \mu_P(i) - \mu_C(i)) + \varepsilon$$

$$\mathrm{Var}[(\hat{Y}_i - Y_i)/Y_i] = \sigma_P^2(i) + \sigma_T^2(i) + \sigma_C^2(i) + \sigma_R^2(i) + \varepsilon$$

and for i in H we have

$$E\left(\frac{\hat{Y}_i - Y_i}{Y_i}\right) = \frac{35}{159Y_i}\left\{\left(\frac{P_i}{P.} + \frac{P_i/C_i}{(P/C).}\right)\mu_P(i) + \frac{U_i}{U.}\ \mu_U(i) - \frac{P_i/C_i}{(P/C).}\ \mu_C(i)\right\} + \frac{27}{159Y_i}\frac{I_i}{I.}\ \mu_I(i) + \varepsilon$$

and

$$\mathrm{Var}\left(\frac{Y_i - Y_i}{Y_i}\right) = \left(\frac{35}{159Y_i}\right)^2\left\{\left(\frac{P_i}{P.} + \frac{P_i/C_i}{(P/C).}\right)^2 \sigma_P^2(i) + \left(\frac{U_i}{U.}\right)^2 \sigma_U^2(i) + \left(\frac{P_i/C_i}{(P/C).}\right)^2 \sigma_C^2(i)\right\}$$

$$+ \left(\frac{27}{159Y_i}\right)^2\left\{\left(\frac{T_i^2/R_i}{(T^2/R).}\right)^2 \left(4\,\sigma_T^2(i) + \sigma_R^2(i)\right) + \left(\frac{I_i}{I.}\right)^2 \sigma_I^2(i)\right\} + \varepsilon \quad .$$

Each of the μ and σ^2 terms in the four formulas above refers to the mean and variance of a state's relative error about the national relative error.

In chapter 6 estimates of $E(\hat{Y}_i - Y_i)$, $\mathrm{Var}(\hat{Y}_i)$, and $E|\hat{Y}_i - Y_i|$ will be given. These estimates are based on formulas (4.20)-(4.23) above.

The remainder of this chapter is technical, devoted to derivations of the above relations, and may be skipped on a first reading.

§4.4 Assumptions and Degrees of Approximation

Let Z denote a generic data element. As discussed in §2.3 above, the delta method will be applied to rational functions of the $\{Z_i\}$ such that the denominator has non-negative coefficients and the function is bounded, e.g. (4.1), (4.2). The analysis of functions such as

$$(4.24) \qquad Y_i = \frac{X_i}{X.} \frac{\max(S_i, H_i)}{\Sigma \max(S_j, H_j)}$$

has been discussed in §2.3. Application of the "modified" delta method will refer to application of the delta method to function like (4.24) or, more generally, (2.16). It follows from the specification of the error models (chapter 3) and the discussion of §2.3 that assumptions (2.3)-(2.6) of the delta method are satisfied.

Not only the observed total national value of each data element, $Z.$, but also the individual state values, Z_i , are bounded random variables (see §2.3). It is hard to say what the ranges of Z_i and $Z.$ random variables are, but for the present purposes we may assume that with probability 1

$$(4.25) \qquad |z./Z.| \leq |E\, z./Z.| + 3[\mathrm{Var}(z./Z.)]^{1/2}$$

and

$$|z_i/Z_i - z./Z.| \leq |\mu_Z(i)| + 3\,\sigma_Z(i)$$

where $\mu_Z(i)$ and $\sigma_Z^2(i)$ are the mean and variance of $z_i/Z_i - z./Z.$.

Applying the delta method to $\hat{Z}_i/\hat{Z}. - Z_i/Z.$ yields a remainder ε_i precisely of the form

$$(4.26) \qquad \varepsilon_i = (\hat{Z}_i/\hat{Z}. - Z_i/Z.) - (Z_i/Z.)(z_i/Z_i - z./Z.)$$

where $(Z_i/Z.)(z_i/Z_i - z./Z.)$ is the main term in the approximation. The ratio of the remainder to the main term simplifies in the following manner:

$$(4.27) \qquad \frac{\varepsilon_i}{\frac{Z_i}{Z.}\left(\frac{z_i}{Z_i} - \frac{z.}{Z.}\right)} = \left\{ \frac{1+z_i/Z_i}{1+z./Z.} - 1 \right\} \left\{ \frac{z_i}{Z_i} - \frac{z.}{Z.} \right\}^{-1} -1$$

$$= \quad (1+z./Z.)^{-1} \; -1 \quad .$$

The very worst state-level data element, i.e. that with the widest possible range of relative error, was state income tax amount I_i = median $\{.01L_i \, , .15K_i \, , .06L_i\}$ where L_i = federal individual income tax liabilities of state i, K_i = state individual income tax collections in state i. Even here (see §3.11.4)

$$|Ei./I.| \leq .063$$

and

$$(\mathrm{Var}(i./I.))^{1/2} \leq .020$$

so with assumption (4.24) , $|i./I.| \leq .123$ with probability 1 and the ratio of remainder to the main term is smaller in absolute value than

$$\frac{1}{1-.123} - 1 = .140 \quad .$$

The bound of .140 is an extreme bound for the ratio of remainder term to the main term in (4.25). Other data elements are better than I_i and yield smaller bounds. It is desirable to obtain bounds for the remainder terms arising from applications of the delta method to more complicated expressions than $\hat{Z}_i/\hat{Z}. - Z_i/Z.$. Preliminary analysis did not provide useful expressions for these bounds.

We next consider the closeness of terms such as $p./P.$, and $\langle p/P\rangle'$ discussed in §3.14.4. In §3.2 the method used to estimate $\sigma_P^2(i) = \mathrm{Var}(p_i/P_i - p./P.)$ would not have yielded a different estimate for $\mathrm{Var}(p_i/P_i - \langle p/P\rangle')$. The reason is the closeness of $p./P.$ and $\langle p/P\rangle'$ relative to the precision underlying the estimation of $\sigma_P^2(i)$. Specifically, it was shown in §3.14.4 that $E(p./P. - \langle p/P\rangle')^2 \leq 9.7 \times 10^{-8}$. The data and assumptions used in §3.2.3,4 to estimate $\sigma_P^2(i)$ make a difference of this magnitude undetectable. Recall from §3.2 that for i = New Jersey $\sigma_P^2(i)$ is between 10^{-4} and 10^{-5} (except for EP's 1 and 2 when

$\sigma_P^2(i) = 0$) and that $\tilde{\mu}_P(i)$ is about 10^{-2}. Similarly, the difference between $E(p./P. - \langle p/P\rangle')$ and $E(p_i/P_i - p./P.) = \mu_P(i)$ is $E(p./P. - \langle p/P\rangle')$, which is smaller in absolute value than 3.11×10^{-4}. The procedures used to estimate $\mu_P(i)$ are not sensitive to such a small difference. In the sequel we will proceed as if

(4.28) $$\mathrm{Cov}\{ p_i/P_i - \langle p/P\rangle' , p_j/P_j - \langle p/P\rangle' \}$$

$$= \mathrm{Cov}\{p_i/P_i - p./P. , p_j/P_j - p./P.\}$$

and

(4.29) $$E(p_i/P_i - \langle p/P\rangle') = E(p_i/P_i - p./P.) \quad .$$

Similar relations will be used for other state-level data elements, e.g. $E(t_i/T_i - \langle t/T\rangle''') = E(t_i/T_i - t./T.)$.

§4.5 Derivations and Proofs

To derive formulas (4.15)-(4.22) the delta method will be applied in stages. This has two advantages:

(i) interesting intermediate results are obtained (e.g. (4.30)-(4.38) below, and

(ii) the problem is broken into smaller, simpler pieces.

Propositions 4.1 and 4.2 are applications of the delta method to the Senate and House formulas. Proposition 4.3 links the results of these two propositions, yielding the conclusions (4.15)-(4.22) as easy consequences.

Proposition 4.1

With the definition of $\hat{S}_i$, S_i , $\mu_S(i)$, $\sigma_S^2(i)$, $\langle p/P\rangle'$ etc. the following properties hold:

$$\text{(4.30)} \qquad \frac{\hat{S}_i - S_i}{S_i} = \left(\frac{p_i}{P_i} - \langle p/P\rangle'\right) + \left(t_i/T_i - \langle t/T\rangle'\right) - \left(c_i/C_i - \langle c/C\rangle'\right) - \left(r_i/R_i - \langle r/R\rangle'\right) + o_p(N^{-1/2})$$

$$\text{(4.31)} \qquad E\left(\frac{\hat{S}_i - S_i}{S_i}\right) = \mu_S(i) + o(N^{-1/2})$$

where $\mu_S(i)$ is defined in (4.8)

$$\text{(4.32)} \qquad \mathrm{Var}(\hat{S}_i/S_i) = \sigma_S^2(i) + o(N^{-1})$$

where $\sigma_S^2(i)$ is defined in (4.9)

and

$$\text{(4.33)} \qquad \operatorname{Cov}(\hat{S}_i/S_i \; , \; \hat{S}_j/S_j) = \sigma_{SS}(i,j) + o(N^{-1}) \qquad i \neq j$$

where $\sigma_{SS}(i,j)$ is defined in (4.12) .

Proof: Let $\underset{\sim}{h} = (h_1,\ldots,h_n)^T$ be defined by

$$h_i(p_1,c_1,t_1,r_1,p_2,c_2,\ldots,p_n,c_n,t_n,r_n) =$$

$$\left\{\frac{(P_i+p_i)(T_i+t_i)(R_i+r_i)^{-1}(C_i+c_i)^{-1}}{\sum_j (P_j+p_j)(T_j+t_j)(R_j+r_j)^{-1}(C_j+c_j)^{-1}}\right\} \cdot \left\{\frac{P_iT_i/(R_iC_i)}{\sum_j P_jT_j/(R_jC_j)}\right\}^{-1} - 1 \; (= \frac{\hat{S}_i - S_i}{S_i}) \; .$$

It is known from discussion in §4.4 and §2.3 that $\underset{\sim}{h}$ satisfies the assumptions of theorem 2.1. It is also known from chapter 3 that the assumptions about orders of magnitudes of the relative errors are satisfied. Recall that N is an indicator for order of magnitude.

For simplicity of presentation consider $i = 1$. The first row of $\nabla \underset{\sim}{h}(\underset{\sim}{0})^T$ in (2.20) is:

$$S_1(\frac{1}{P_1} - \frac{S_1}{P_1} \, , \, \frac{-1}{C_1} + \frac{S_1}{C_1} \, , \, \frac{1}{T_1} - \frac{S_1}{T_1} \, , \, \frac{-1}{R_1} + \frac{S_1}{R_1} \, , \, -\frac{S_2}{P_2} \, ,$$

$$+\frac{S_2}{C_2} \, , \, -\frac{S_2}{T_2} \, , \, +\frac{S_2}{R_2} \, , \, -\frac{S_3}{P_3} \, , \, +\frac{S_3}{C_3} \, ,\ldots, \frac{S_n}{R_n}) \; .$$

Thus, $h_1(p_1,c_1,\ldots,t_n,r_n) = \frac{p_1}{P_1}(1-S_1) + \frac{c_1}{C_1}(-1+S_1)$

$$+ \frac{t_1}{T_1}(1-S_1) + \frac{r_1}{R_1}(-1+S_1)$$

$$+ \sum_{j=2}^{n} \left\{ - S_j \frac{p_j}{P_j} + S_j \frac{c_j}{C_j} \right.$$

$$\left. - S_j \frac{t_j}{T_j} + S_j \frac{r_j}{R_j} \right\} + o_p(N^{-1/2})$$

$$= (p_1/P_1 - \sum_j S_j p_j/P_j) + (t_1/T_1 - \sum_j S_j t_j/T_j)$$

$$- (c_1/C_1 - \sum_j S_j c_j/C_j)$$

$$- (r_i/R_i - \sum_j S_j r_j/R_j) \qquad + o(N^{-1/2})$$

which becomes (4.30) when the definitions of the weighted sums $\langle p/P\rangle'$, $\langle c/C\rangle'$, etc. are applied.

Formula (4.31) can be obtained either by a direct application of theorem 2.1 or in a heuristic manner by taking the expectation of (4.30). Notice that $\mu_S(i)$ (defined in (4.8)) uses the assumption that the estimates of net state and local taxes and personal income are unbiased (see §3.8.5, §3.7.5). Assumptions like (4.29) are also used.

Formula (4.32) is obtained either in the same way as corollary 2.2 or heuristically as a variance of the main term in (4.30). Notice that $\sigma_S^2(i)$ (defined in (4.9)) utilizes the assumption that the estimates of population, per capita income, personal income, and net state and local taxes are uncorrelated (see §3.14.4).

Formula (4.33) is an immediate consequence of corollary 2.2. Notice that $\sigma_S^2(i,j)$ is defined by (4.12), which utilizes (4.4). This uses the assumption that the data elements are uncorrelated between states as well as within states (see §3.14.4). Assumptions like (4.28) are also used.

*****

Proposition 4.2

With the definitions of $\hat{H}_i$, H_i , $\mu_H(i)$, etc. the following definitions hold:

(4.34)
$$\frac{\hat{H}_i - H_i}{H_i} = \frac{35}{159} H_i^{-1} \left\{ \frac{P_i}{P.}\left(\frac{p_i}{P_i} - \frac{p.}{P.}\right) + \frac{P_i/C_i}{(P/C).}\left(\frac{p_i}{P_i} - \langle p/P\rangle''\right) + \frac{U_i}{U.}\left(\frac{u_i}{U_i} - \frac{u.}{U.}\right) - \frac{P_i/C_i}{(P/C).}\left(\frac{c_i}{C_i} - \langle c/C\rangle''\right) \right\}$$
$$+ \frac{27}{159} H_i^{-1} \left\{ \frac{2T_i^{\,2}/R_i.}{(T^2/R).}\left(\frac{t_i}{T_i} - \langle t/T\rangle'''\right) + \frac{I_i}{I.}\left(\frac{i_i}{I_i} - \frac{i.}{I.}\right) - \frac{T_i^{\,2}/R_i}{(T^2/R).}\left(\frac{r_i}{R_i} - \langle r/R\rangle'''\right) \right\} + o_p(N^{-1/2})$$

(4.35)
$$E\left(\frac{\hat{H}_i - H_i}{H_i}\right) = \mu_H(i) + o(N^{-1/2})$$

where $\mu_H(i)$ is defined in (4.10)

(4.36) $\mathrm{Var}(\hat{H}_i/H_i) = \sigma_H^2(i) + o(N^{-1})$

where $\sigma_H^2(i)$ is defined in (4.11)

(4.37) $\mathrm{Cov}(\hat{H}_i/H_i , \hat{H}_j/H_j) = \sigma_{HH}(i,j) + o(N^{-1}) \quad i \neq j$

where $\sigma_{HH}(i,j)$ is defined in (4.14)

(4.38) $\mathrm{Cov}(\hat{H}_j/H_j , \hat{S}_i/S_i) = \sigma_{SH}(i,j) + o(N^{-1}) \quad i \neq j$

where $\sigma_{SH}(i,j)$ is defined in (4.13) .

Proof: The proof follows along the lines of proposition 4.1 except that additional absences of correlation are utilized (§3.14.4).

*****

Recall $X_i = \max(S_i , H_i)$, $\hat{X}_i = \max(\hat{S}_i , \hat{H}_i)$. The next proposition derives approximations for $\hat{X}_i/\hat{X}. - X_i/X.$ in terms of of the approximations for $(\hat{S}_i - S_i)/S_i$ and $(\hat{H}_i - H_i)/H_i$ obtained by the two preceding propositions. The proof is also a straightforward application of the modified delta method.

Proposition 4.3

Let $x_i = \hat{X}_i - X_i$. Then with the definitions of σ_{SS} , σ_{SH} , σ_S^2 etc. the following relations hold:

(4.39) $$\frac{\hat{X}_i/\hat{X}. - X_i/X.}{X_i/X.} = x_i/X_i - x./X. + o_p(N^{-1/2})$$

(4.40) $$E(\frac{\hat{X}_i/\hat{X}. - X_i/X.}{X_i/X.}) = E(x_i/X_i - X./X.) + o(N^{-1/2})$$

(4.41) $$\mathrm{Var}(\frac{\hat{X}_i/\hat{X}.}{X_i/X.}) = \mathrm{Var}(\frac{\hat{X}_i}{X_i}) - 2\sum_j \frac{X_j}{X.}\mathrm{Cov}(\frac{\hat{X}_i}{X_i}, \frac{\hat{X}_j}{X.})$$

$$+ \sum_{j,k} \frac{X_j}{X.}\frac{X_k}{X.} \mathrm{Cov}(\frac{\hat{X}_j}{X_j}, \frac{\hat{X}_k}{X_k}) + o(N^{-1})$$

where $x_i/X_i = \frac{\hat{H}_i - H_i}{H_i} + o_p(N^{-1/2})$ for i in H

$= \frac{\hat{S}_i - S_i}{S_i} + o_p(N^{-1/2})$ for i in S

and

$E(x_i/X_i) = \mu_H(i) + o(N^{-1/2})$ for i in H

$= \mu_S(i) + o(N^{-1/2})$ for i in S

where $\mu_H(i)$ and $\mu_S(i)$ are defined in (4.10) and (4.8) .

and

$$\operatorname{Cov}\left(\frac{\hat{X}_i}{X_i}, \frac{\hat{X}_i}{X_i}\right) = \operatorname{Var}\left(\frac{\hat{X}_i}{X_i}\right) = \sigma_H^2(i) + o(N^{-1}) \qquad \text{if } i \text{ is in } H$$

$$= \operatorname{Var}\left(\frac{\hat{X}_i}{X_i}\right) = \sigma_S^2(i) + o(N^{-1}) \qquad \text{if } i \text{ is in } S$$

where $\sigma_H^2(i)$ and $\sigma_S^2(i)$ are defined in (4.11) and (4.9)

and

$$\operatorname{Cov}\left(\frac{\hat{X}_i}{X_i}, \frac{\hat{X}_j}{X_j}\right) = \sigma_{HH}(i,j) + o(N^{-1}) \qquad \text{if } i \text{ and } j \text{ are in } H,\ i \neq j$$

$$= \sigma_{SH}(i,j) + o(N^{-1}) \qquad \text{if } i \text{ is in } S \text{ and } j \text{ is in } H$$

$$= \sigma_{SS}(i,j) + o(N^{-1}) \qquad \text{if } i \text{ and } j \text{ are in } S,\ i \neq j$$

where $\sigma_{HH}(i,j)$, $\sigma_{SH}(i,j)$, and $\sigma_{SS}(i,j)$ are defined in (4.12), (4.13), and (4.14).

Proof: Formula (4.39) results from application of the modified delta method (see §2.3) and formulas (4.30) and (4.31). Notice that $(\hat{X}_i - X_i)/X_i$ is approximated by $(\hat{S}_i - S_i)/S_i$ for i in S and $(\hat{H}_i - H_i)/H_i$ for i in H . These kinds of approximations were discussed in §2.3. Formulas (4.40) and (4.41) can be obtained either the same way as (4.39) or by heuristic manipulation of (4.39).

*****

Conclusions (4.17), (4.20) and (4.21) follow easily by substituting in (4.39)-(4.41). For example if i is in $\mathcal{S}$ then (4.39) becomes

$$\frac{\hat{Y}_i - Y_i}{Y_i} = \frac{\hat{S}_i - S_i}{S_i} - \frac{\sum_{\mathcal{S}} (\hat{S}_j - S_j) + \sum_{\mathcal{H}} (\hat{H}_j - H_j)}{\sum_{\mathcal{S}} S_j + \sum_{\mathcal{H}} H_j} + o_p(N^{-1/2}) \quad .$$

Similarly, for i in $\mathcal{H}$ (4.39) becomes

$$\frac{\hat{Y}_i - Y_i}{Y_i} = \frac{\hat{H}_i - H_i}{H_i} - \left(\frac{\sum_{\mathcal{S}} (\hat{S}_j - S_j) + \sum_{\mathcal{H}} (\hat{H}_j - H_j)}{\sum_{\mathcal{S}} S_j + \sum_{\mathcal{H}} H_j} \right) + o_p(N^{-1/2}) \quad ,$$

and (4.17) is immediate. Formulas (4.20) and (4.21) follow from substitution of Y_j for X_j/X. Finally, (4.23) follows by substituting $\mu_Y(i)$, $\sigma_Y^2(i)$ in (2.28).

Although the present analysis is univariate and we do not need the covariance between allocations to different states, approximations to the covariances are easily obtained by the methods used above.

Chapter 5 Intrastate Allocations in GRS

§5.1 Overview

This chapter analyzes how errors in data affect the distribution of GRS funds below the state level. The substate formulas are described in great detail (§5.2), and notation (§5.3) is developed to permit rigorous analysis of the approximate means, variances, and covariances of errors in the substate allocations. Analysis is complicated because the substate allocations must be determined by an iterative algorithm. Since iterative allocation formulas are not uncommon the analytic approach taken should be useful for studying other allocation programs as well.

The reader is encouraged to read §5.2 in order to get some feeling for the complexity of the formula. By following the examples there the reader can actually understand how the formula determines the allocations. For a simple overview see Appendix B. The reader interested primarily in benefit-cost applications and not in GRS *per se* may skip §§5.3-5.4 and proceed from §5.2 directly to chapter 6. Explicit references to results in §§5.3-5.5 will be made there so that the reader can easily refer back if the desire arises.

§5.2 Determination of Substate Allocations

The distribution of a state's allocation to units within the state is determined by a complex procedure. To ensure that no one gets too much or too little, a myriad of floors and ceilings affect the allocations. The precise procedure that determines the allocations was not specified in the legislation, but was formulated by the Department of the Treasury (see Joint Committee on Internal Revenue Taxation 1973, p. 34). As far as this researcher could discover, the only publicly available complete description of the allocation procedure is the computer program and documentation written by Westat, Inc. In its most general form, the allocation procedure permits states to choose optional formulas. Special provisions also apply to Alaskan native villages and to Indian tribes. The allocation procedure used for New Jersey was not complicated by these special considerations and is described below.

After the allocations to the 51 states are determined, each state's allocation is partitioned among various governments within the state according to a doubly iterative algorithm, wherein an "inner" iterative procedure is nested within an "outer" iterative procedure. The rth round of the outer iteration (initially $r = 1$) begins with the partitioning of the "local share" $Y_{L(r)}$ (where $Y_{L(1)} = \frac{2}{3}Y$) among the geographic <u>county areas</u> proportionally by the product of tax effort (see §3.14) and population divided by per capita income.* If any county area's share, on a per capita basis, would exceed 145% of the ratio of $(\frac{2}{3})Y$ to state population ("145% PCLS** limit"), its share is reduced to that level and the excesses distributed to each county area not at the 145% PCLS limit, in proportion to the county area's share. Iteration may be necessary to ensure that no county area share exceeds the 145% PCLS limit. Next, if any county area's share, on a per capita basis, would fall below 20% of the ratio of $(\frac{2}{3})Y$ to total state population ("20% PCLS limit"), its share is increased to that level by taking from each county area whose share is between the 20% and 145% PCLS limits, in proportion to the county area's share. Again, iteration may be necessary to ensure that no county area share falls below the 20% PCLS limit. The application of the PCLS limits constitutes the "inner" iteration alluded to above.

The inner iteration process mentioned above nests within an outer iteration process as follows. A county area's share is allocated to three groups within the county area, namely the county government,

---

* During any single round of the outer iteration, numerous iterations may be performed to partition the local share among county areas. These latter iterations are referred to as "inner iterations". As will be explained below, the outer iterations determine the size of the local share.

** Per Capita Local Share

all townships, and all "places and municipalities" (hereafter, "places") proportionally by adjusted taxes (see §3.12). The amount to all townships is then partitioned among all townships in the county area proportionally by the product of tax effort and population divided by per capita income. If any township share exceeds on a per capita basis the 145% PCLS limit (described above), the share is reduced to that level and the excess is "set aside". If the share to a township falls below the 20% PCLS limit on a per capita basis, it is increased to the smaller of the following two levels: (i) the 20% PCLS limit on a per capita basis, or (ii) one half (or one quarter for a six month EP) the sum of the township's intergovernmental transfers of revenue (see §3.13) and adjusted taxes, on total basis ("50% ceiling"). The deficit is then "set aside". At this point any township allocation greater than the 50% ceiling is reduced to that level and the excess transferred to the county government. A similar procedure is then followed for places.

The net amount "set aside" (excesses minus deficits), say $Y_{\Delta(r)}$, from application of the 145% PCLS, 20% PCLS constraints on subcounty units is computed. If the actual amount allocated differs from $\frac{2}{3}Y$ by less than \$3, i.e. if

$$(5.1) \qquad |Y_{L(r)} - Y_{\Delta(r)} - \tfrac{2}{3}Y| \leq \$3 ,$$

then the iteration procedure is finished. Otherwise the whole process is begun again, except that instead of using $Y_{L(r)}$ for the local share, the local share is set at $Y_{L(r+1)}$, given by

$$(5.2) \qquad Y_{L(r+1)} = \frac{Y_{L(r)} \frac{2}{3}Y}{Y_{L(r)} - Y_{\Delta(r)}} .$$

It is worth noting that the actual algorithm is even more complicated than described above. For clarity a simpler but algebraically identical algorithm has been described. The differences between the two algorithms pertain to the way that $Y_{L(r+1)}$ is computed.

Suppose that (5.1) has been satisfied, so that the iterations have terminated. If the share to a county government exceeds its 50% ceiling, the share is reduced to that level and the excess is given to the state government. Next, any place or township allocation less than \$200 (or \$100 for a six month EP; this is called de minimis) is forfeited by the unit and given to the county government, unless the county government was at the 50% ceiling, in which case the forfeitures go to the state government. Occasionally a local government waives its allocation, and these waived amounts are treated identically to forfeited amounts.

After the final allocations to substate units have been determined, the state government allocation is calculated. Careful analysis of the above scenario shows that the state government allocation equals $Y/3$ plus all excesses from applications of the 50% ceilings to county governments plus all waived and forfeited amounts from units in county areas where the county government was at the 50% ceiling.

The following example illustrates the algorithm in a simple, artificial situation.

Example 5.1

Suppose population of the state is 10^4 people and the allocation to the whole state is $Y = \$3 \times 10^8$. Then

$$Y_{L(1)} = (2Y/3) = 2 \times 10^8$$

$$20\%\ \text{PCLS} = (.2)(2 \times 10^8) \times 10^{-4} = 4 \times 10^3$$

$$145\%\ \text{PCLS} = (1.45)(2 \times 10^8) \times 10^{-4} = 2.9 \times 10^4 .$$

We will let round(r,s) denote the rth outer iteration, s^{th} inner iteration. The county data is

	County Area 1	County Area 2	County Area 3	Total
Population	10^3	4×10^3	5×10^3	10^4
Per Capita Income	4×10^3	4×10^3	4×10^3	
Adjusted Taxes	3.2×10^7	3.2×10^7	1.6×10^7	
Tax Effort	8	2	8×10^{-1}	
$\frac{\text{Tax Effort} \times \text{Pop.}}{\text{Per Capita Income}}$	2	2	1	5 .

The round (1,1) allocations to county areas are proportional to tax effort × population/percapita income, so

Round(1,1) Alloc.	8×10^7	8×10^7	4×10^7	2×10^8
Round(1,1) Per Capita Alloc.	8×10^4 (>145% PCLS)	2×10^4	8×10^3 .	

The round(1,1) per capita allocation to county area 1 exceeds 2.9×10^4 = 145% PCLS, so its per capita allocation is reduced to 2.9×10^4. The round(1,2) allocation to county area 1 is therefore $10^3 \times 2.9 \times 10^4 = 2.9 \times 10^7$, leaving $8 \times 10^7 - 2.9 \times 10^7 = 5.1 \times 10^7$ to be shared between county areas 2 and 3 (which are between the 20% and 145% PCLS limits) in proportion to their respective products of tax effort and population divided by per capita income. The additional allocation to county area 2 is therefore

$$\frac{8 \times 10^7}{8 \times 10^7 + 4 \times 10^7} \times (5.1 \times 10^7) = 3.4 \times 10^7$$

and the additional allocation to county area 3 is 1.7×10^7.

	County Area 1	County Area 2	County Area 3	Total
Round(1,2) Alloc.	2.9×10^7	11.4×10^7	5.7×10^7	2×10^8
Round(1,2) Per Capita Alloc.	2.9×10^4	2.85×10^4	1.14×10^4	

Each round(1,2) county area allocation is within the PCLS limits, so subcounty allocations are next to be determined. Notice, however that had the ratio of the round(1,1) allocation of county area 2 to that of county area 3 been much larger, the round(1,2) per capita allocation of county area 2 would have exceeded 2.9×10^4, necessitating another round of iteration.

The subcounty allocation will be illustrated for county area 1 only. It is assumed that allocations to governments within county areas 2 and 3 always lie between constraining floors and ceilings. County 1 consists of a county government (CG), 3 places (P1, P2, P3) and one township (T1). The subcounty data is as follows.

	CG	P1	P2	P3	T1	Total
Population		10^2	3×10^2	5×10^2	10^2	10^3
Per Capita Income		2×10^3	4×10^3	4×10^3	6×10^3	
Adjusted Taxes	2.4×10^7	10^6	2×10^6	3×10^6	2×10^6	3.2×10^7
Tax Effort		5	1.67	1.5	3.33	
$\frac{\text{Tax Effort} \times \text{Pop.}}{\text{Per Capita Income}}$		2.5×10^{-1}	1.25×10^{-1}	1.875×10^{-1}	5.56×10^{-2}	
Intergov. Transfers	2.0×10^7	3×10^6	1.8×10^7	3.7×10^7	6.8×10^7	
.5(Adj. Tax + Intergov. Transfer)	2.2×10^7	2×10^6	1×10^7	2×10^7	3.5×10^7	

Note that .5(Adj. Tax + Integov. Transfer) $\geq$ 20% PCLS for all places and townships.

The allocation to county area 1 is first portioned according to adjusted taxes into the county government share, the share to all places, and the share to all townships. Then the share to all places is divided among the places according to the product of tax effort and population divided by per capita income. (There is only one township so it receives the entire township share.) Thus, e.g. the share to the county government is

$$(2.9 \times 10^7) \times (2.4 \times 10^7)/(3.2 \times 10^7) = 2.175 \times 10^7$$

and the share to all places is

$$(2.9 \times 10^6) \times (6 \times 10^7)/(3.2 \times 10^7) = 5.4375 \times 10^6 \quad .$$

The share to all places is split by P1, P2, P3 in the ratio 2.5: 1.25: 1.875. The following display illustrates the results.

	CG	P1	P2	P3	T1	Total
Local Alloc. (r=1)	2.175×10^7	all places = 5.44×10^6			1.81×10^6	2.9×10^7
		2.42×10^6	1.21×10^6	1.81×10^6		
Per Capita Local Alloc. (r=1)		2.42×10^4	4.03×10^3	3.62×10^3 (<20% PCLS)	1.81×10^4	

The allocation to place 3 falls below the 20% PCLS of 4×10^3 so its allocation is increased to $(5 \times 10^2) \times (4 \times 10^3) = 2.0 \times 10^6$ and the "deficit" of $2.0 \times 10^6 - 1.81 \times 10^6 = 1.9 \times 10^5$ is "set aside". Since all other subcounty units in all county areas are between PCLS floors and celings , $Y_{\Delta(1)} = -1.9 \times 10^5$.

Since $Y_{L(1)} = \frac{2}{3}Y = 2 \times 10^8$:

$$|Y_{L(1)} - Y_{\Delta(1)} - \frac{2}{3}Y| = 1.9 \times 10^5 > 3$$

and another round of iteration (r = 2) is begun, using

$$Y_{L(2)} = \frac{(2 \times 10^8)(2 \times 10^8)}{2.0019 \times 10^8} = 1.9981 \times 10^8 \quad .$$

Note that the 20% PCLS, 145% PCLS limits do not change. The new county area allocations are

	County Area 1	County Area 2	County Area 3	Total
Round(2,1) Alloc.	7.992×10^7	7.992×10^7	3.996×10^7	1.998×10^8
Round(2,1) Per Capita Alloc.	7.992×10^4 (>145% PCLS)	1.998×10^4	7.992×10^3	

The county area 1 allocation exceeds the 145% PCLS limit so it is reduced and the excess of $(7.992 \times 10^4 - 2.900 \times 10^4) \times 10^3 = 5.092 \times 10^7$ is split by county areas 2 and 3 in proportion to their respective products of tax effort and population divided by per capita income:

	County Area 1	County Area 2	County Area 3	Total
Round(2,2) Alloc.	2.9×10^7	11.387×10^7	5.693×10^7	1.998×10^8
Round(2,2) Per Capita Alloc.	2.9×10^4	2.847×10^4	1.139×10^4	.

As at iteration $r = 1$, the allocation to county area 1 is split among the county government, all places, and all townships according to adjusted taxes. Then, the share to all places is split by P1, P2, P3 in proportion to tax effort times population divided by per capita income. Since the allocation to county area 1 is the same as at round(1,2), the allocations (before PCLS limits are checked) within county area 1 are also the same. The allocation to P3 is again raised to the 20% PCLS level, and the deficit "set aside":

	CG	P1	P2	P3	T1	Total
Local Alloc. (r = 2)	2.175×10^7	2.42×10^6	1.21×10^6	1.81×10^6	1.81×10^6	2.9×10^7
Per Capita Local Alloc. (r = 2)		2.42×10^4	4.03×10^3	3.62×10^3 (<20% PCLS)	1.81×10^4	
PCLS-Adjusted Per Capita Local Alloc. (r = 2)		2.42×10^4	4.03×10^3	4.00×10^3	1.81×10^4	
PCLS-Adjusted Local Alloc. (r = 2)	2.175×10^7	2.42×10^6	1.21×10^6	2.00×10^6	1.81×10^6	2.919×10^7

Here the deficit is 1.9×10^5 and so $Y_{\Delta(2)} = -1.9 \times 10^5$. However,

$$Y_{L(2)} = 1.9981 \times 10^8$$

so $|Y_{L(2)} - Y_{\Delta(2)} - \frac{2}{3}Y| = |1.9981 \times 10^8 + 1.9 \times 10^5 - 2 \times 10^8| = 0 \leq 3$

and the iterations stop.

The allocation to place P1 exceeds the 50% ceiling however, since $2.42 \times 10^6 > 2.00 \times 10^6$ so the allocation is reduced to 2.00×10^6 and the difference of 4.2×10^5 goes to the county government. The allocation to the county government is now 2.217×10^7, which is greater than its 50% ceiling of 2.2×10^7, so the county government is set at 2.2×10^7 and the remaining 1.7×10^5 goes to the state government. The final allocations within the county area 1 are thus

	CG	P1	P2	P3	T1	Total
Final Alloc.	2.2×10^7	2×10^6	1.21×10^6	2×10^6	1.81×10^6	2.902×10^7

Since no allocation is less than \$200, the de minimis rule does not have an effect. If any subcounty allocation in county area 1 were waived, the amount would go to the state government because county government CG was affected by the 50% ceiling. If CG was not at the 50% ceiling, it would receive any waived amounts from units in county area 1. Assuming no unit waives its allocation, the allocation to the state government is $\frac{1}{3}Y + 1.7 \times 10^5 = 1.0017 \times 10^8$.

To get some feeling for the effect of the constraints in New Jersey note that in EP 1

$$\frac{\text{final } Y_{L(r)}}{\frac{2}{3}Y} = 1.027 \quad ,$$

and the following table holds.

	County Area	County Gov.	Places	Townships
Total No. in N.J.	21	21	334	232
No. Constrained by 145% PCLS	4	-	24	7
No. Constrained by 20% PCLS	0	-	29	23
No. Constrained by 50% Ceiling	-	0	0	0
No. Waived or Forfeited	-	0	0	0

Table 5.0

Substate Units Constrained in New Jersey, EP 1

The remainder of this chapter involves massive detail. The reader should feel free to merely skim the rest of this chapter or to skip to chapter 6 and refer back as need arises.

§5.3 Notation

The state of reference (New Jersey) is fixed and we will no longer use the subscript i for states. The subscript j will refer to counties and k will refer to subcounty units. The collections of subscripts j,k are denoted by $\mathcal{J}$, $\mathcal{K}$. There are two types of subcounty governments in New Jersey, townships and places, and the set $\mathcal{K}$ is partitioned accordingly into $\mathcal{K}1$ and $\mathcal{K}2$, with elements k1 and k2.

As usual, we denote parameter values (estimates) by capital letters (with a "^"). Only notation for parameter values will be explicitly defined; notation for estimates values is implicit. The symbols Y, Y_j, $Y_{jk\alpha}$, represent the allocations to the whole state (New Jersey), to county government j, and to local government $jk\alpha$. The final values in the iteration procedure of $Y_{L(r)}$, $Y_{\Delta(r)}$, are denoted by Y_L, Y_Δ.

The following table will facilitate identification of various substate data elements and error components.

Symbol	Data Element	Section for Location of Definition
P, σ_P^2, μ_P	population	§3.4
C, σ_C^2, μ_C	per capita income	§3.6
D, σ_D^2, μ_D	adjusted taxes	§3.12
G, σ_G^2, μ_G	intergovernmental transfers	§3.13

Table 5.1

Data Element References

Define

(5.3) $$X_j = \frac{(D_j + D_j.)/C_j^2}{\sum_{\ell \varepsilon J} (D_\ell + D_\ell.)/C_\ell^2}$$

and note that X_j is the proportion of $Y_{L(r)}$ allocated to county area j at round(r,1).

The following quantity is useful for representing the final substate allocations by algebraic formulas. Define

V = ratio of

(1) allocation at final round of outer iteration to a county area not at a PCLS limit after all county area allocations greater than 145% PCLS limits have been reduced and the excesses distributed and after all county area allocations less than the 20% PCLS limits have been increased by taking from all county areas not at a 20% or 145% PCLS limit

to

(2) allocation to the county area at round(1,1) before application of any PCLS constraints.

Note that (1) above can be written as

(5.4) $$X_j Y_L + X_j \gamma$$

where γ denotes the amount by which the allocation $X_j Y_L$ (at final outer iteration, first inner iteration) was increased (or decreased) by distributing excesses and deficits from 145% PCLS and 20% PCLS constrained county areas. Recall that these excesses and deficits were shared in proportion to the tax effort times population divided by per capita income, hence in proportion to X_j. Thus γ is constant for all unconstrained county areas j . Observing that (2) above equals

(5.5) $$X_jY_{L(0)} = \frac{2}{3}X_jY$$

we obtain

$$V = \frac{Y_L + \gamma}{\frac{2}{3}Y} \quad .$$

Notice that V is constant for all counties. It does not depend on j.

To calculate V for the example in §5.1 we consider county area 2 (we could have used county area 3). Here (1) equals 11.387×10^7 (the allocation to county area 2 at round(2,2)) and (2) equals 8×10^7 (the allocation to county area 2 at round(1,1)) so $V = 11.387/8 = 1.4234$.

Notice that the definition of V and (5.5) together imply that the final allocation to a county area j not at a PCLS limit is

$$\frac{2}{3}YVX_j$$

so the share of the substate pie to county area j is VX_j.

Define

(5.6) $$X_{jk\alpha} = \left(\frac{\sum_{\ell\in K\alpha} D_{j\ell\alpha}}{D_j + D_{j\cdot}}\right)\left(\frac{D_{jk\alpha}/C^2_{jk\alpha}}{\sum_{\ell\in K\alpha} D_{j\ell\alpha}/C^2_{j\ell\alpha}}\right)$$

and observe that $X_{jk\alpha}$ is the portion of the county area j allocation which subcounty area $jk\alpha$ gets before subcounty PCLS limits are checked.

Next, define

$\hat{O}_j$ = the amount ultimately received by the government of county j from places and townships constrained as a result of the 50% ceilings and from places and townships waiving or forfeiting their amounts.

Table 5.2 below gives explicit formulas for the allocations to substate units. These formulas follow immediately from the notation for $Y, V, X_j, X_{jk\alpha}$, $\hat{O}_j$.

Type of Unit	Operating Constraint: County Area	Operating Constraint: Local	Allocation Formula *
County Area(j)	None	---	$\frac{2}{3}$ Y V X_j
	PCLS	---	$\frac{2}{3}$ Y $\begin{Bmatrix}1.45\\ .20\end{Bmatrix}$ $(P_j/P.)$
County Government (j)	None	None	$\frac{2}{3}$ Y V X_j $(D_j/(D_j{+}D_j.)) + O_j$
	PCLS	None	$\frac{2}{3}$ Y $\begin{Bmatrix}1.45\\ .20\end{Bmatrix}$ $(P_j/P.)(D_j/(D_j{+}D_j.)){+}O_j$
		50% Ceiling	$.5\ (D_j + G_j)$ (or $.25(D_j + G_j)$ for a 6 month EP)
Township (α = 1) or Place (α = 2)	None	None	$\frac{2}{3}$ Y V X_j $X_{jk\alpha}$
	PCLS	None	$\frac{2}{3}$ Y $\begin{Bmatrix}1.45\\ .20\end{Bmatrix}$ $(P_j/P.)$ $X_{jk\alpha}$
		PCLS	$\frac{2}{3}$ Y $\begin{Bmatrix}1.45\\ .20\end{Bmatrix}$ $P_{jk\alpha}/P..$
		50% Ceiling	$.5\ (D_{jk\alpha}{+}G_{jk\alpha})$ (or $.25(D_{jk\alpha}{+}G_{jk\alpha})$ for a 6 month EP)
		De Minimis or Waived	0

Table 5.2 Substate Allocation Formulas
(adapted from Stanford Research Institute 1974b)

* Notation $\begin{Bmatrix}1.45\\ .20\end{Bmatrix}$ means the appropriate constant should be selected.

Recall that the variances σ_C^2 , σ_D^2 , and covariances σ_{CC} have been defined in §3.6.1,§3.12.1, and §3.6.1 and the weights w_j , $w_{jk\alpha}$ are given by (3.54), (3.55). Also recall that μ_Y , σ_Y^2 , and $\hat{Y}$-Y may be evaluated according to (4.17)-(4.19). Now define

(5.7) $$\sigma_\xi^2(j) = [(1-w_j)/(D_j+D_j.)]^2 \; [D_j^2 \; \sigma_D^2(j) + \sum_{k\varepsilon K} D_{jk}^2 \; \sigma_D^2(j,k)] + \sum_{\substack{\ell\neq j\\ \ell\varepsilon J}} [(w_\ell/(D_\ell+D_{\ell.}))^2 \; (D_\ell^2 \; \sigma_D^2(\ell) + \sum_{k\varepsilon K} D_{\ell k}^2 \; \sigma_D^2(\ell,k)] + 4 \; \sigma_C^2(j)$$

(5.8) $$\sigma_{DD}^2(j) = (D_j+D_j.)^{-2} \; [D_j^2. \; \sigma_D^2(j) + \sum_{k\varepsilon K} D_{jk}^2 \; \sigma_D^2(j,k)]$$

(5.9) $$\sigma_\xi^2(j,k\alpha) = (1-2w_{jk\alpha}) \; (\sigma_D^2(j,k\alpha) + 4 \; \sigma_C^2(j,k\alpha)) + \sum_{\ell\varepsilon K\alpha} w_{j\ell\alpha}^2 \; (\sigma_D^2(j,\ell\alpha) + 4 \; \sigma_C^2(j,\ell\alpha)) - 8[\sum_{\substack{\ell\varepsilon K\alpha\\ \ell\neq k}} w_{j\ell\alpha} \; \sigma_{CC}(j,k\alpha,\ell\alpha)] + 4[\sum_{\substack{\ell,m\varepsilon K\alpha\\ \ell\neq m}} w_{j\ell\alpha} \; w_{jm\alpha} \; \sigma_{CC}(j,\ell\alpha,m\alpha)]$$

and

(5.10) $$\sigma_{\phi}^{2}(j,\alpha) = (D_j + D_{j}.)^{-2} \{ [\sum_{k\varepsilon K\alpha} D_{jk\alpha}^{2} \sigma_{D}^{2}(j,k\alpha)] \cdot [(D_j + \sum_{\ell\varepsilon K\alpha'} D_{j\ell\alpha'})/ \sum_{\ell\varepsilon K\alpha} D_{j\ell\alpha}]^{2} + D_j^{2} \sigma_{D}^{2}(j) + \sum_{\ell\varepsilon K\alpha'} D_{j\ell\alpha'}^{2} \sigma_{D}^{2}(j,\ell\alpha') \}$$

where $\alpha' = 1$ if $\alpha = 2$ and $\alpha' = 2$ if $\alpha = 1$.

The quantities $\sigma_{\xi}^{2}(j)$, $\sigma_{DD}^{2}(j)$, $\sigma_{\xi}^{2}(j,k\alpha)$, $\sigma_{\phi}^{2}(j,\alpha)$ just defined are approximations to the variances of ξ_j , $d_j/D_j - (d_j + d_{j}.)/(D_j + D_{j}.)$, $\xi_{jk\alpha}$, and $\phi_{j\alpha}$ where ξ_j , $\xi_{jk\alpha}$, and $\phi_{j\alpha}$ are defined by (3.60)-(3.62). Also define

(5.11) $$\sigma_{\xi\xi}(j,k\alpha) = (1-w_j)/(D_j + D_{j}.) [D_{jk\alpha}\sigma_{D}^{2}(j,k\alpha) - \sum_{\ell\varepsilon K\alpha} w_{j\ell\alpha} D_{j\ell\alpha} \sigma_{D}^{2}(j,\ell\alpha)]$$

(5.12) $$\sigma_{\xi\phi}(j,\alpha) = -(1-w_j)/(D_j + D_{j}.) [(D_j^{2} \sigma_{D}^{2}(j) + \sum_{k\varepsilon K} D_{jk}^{2} \sigma_{D}^{2}(j,k))/(D_j + D_{j}.) -(\sum_{k\varepsilon K\alpha} D_{jk\alpha}^{2} \sigma_{D}^{2}(j,k\alpha))/(\sum_{k\varepsilon K\alpha} D_{jk\alpha})]$$

and

$$(5.13)\qquad \sigma'_{\xi\phi}(j,k\alpha) = [(\sum_{\ell\varepsilon K\alpha} D_{j\ell\alpha})^{-1} - (D_j + D_{j\cdot})^{-1}]$$

$$\cdot [D_{jk\alpha}\sigma_D^2(j,k\alpha) - \sum_{\ell\varepsilon K\alpha} w_{j\ell\alpha} D_{j\ell\alpha}\sigma_D^2(j,\ell\alpha)] \quad .$$

The quantities $\sigma_{\xi\xi}(j,k\alpha)$, $\sigma_{\xi\phi}(j,\alpha)$, and $\sigma'_{\xi\phi}(j,k\alpha)$ are approximations to $\mathrm{Cov}(\xi_j , \xi_{jk\alpha})$, $\mathrm{Cov}(\xi_j , \phi_{j\alpha})$, and $\mathrm{Cov}(\xi_{jk\alpha} , \phi_{j\alpha})$. The notation just defined by (5.7)-(5.13) is useful for evaluating the variances of relative errors in allocations (§5.5). To evaluate the mean of $\xi_{jk\alpha}$, the quantity $\mu_\xi(j,k\alpha)$ defined by

$$(5.14)\qquad \mu_\xi(j,k\alpha) = -2(\mu_C(j,k\alpha) - \sum_{\ell\varepsilon K\alpha} w_{j\ell\alpha}\, \mu_C(j,\ell\alpha))$$

will be used.

§5.4 Assumptions and Degrees of Approximation

Lacking an implicit or explicit representation for V we cannot apply the delta method to $(\hat{V}-V)/V$. Calculations show that in New Jersey $\hat{V}$ varied from 1.116 for EP's 1-3 to 1.054 for EP 6. It is hypothesized that the variation in the New Jersey data elements across these entitlement periods is greater than the probable range due to error within any one entitlement period. This hypothesis seems reasonable but has not been tested. Note that the biases in allocation, which are caused largely by biases in the decennial census estimates of population and per capita income, will not vary much over time and will not cause variation in $\hat{V}$ over time, although they do contribute to $\hat{V}-V$. If the hypothesis is true, then it is reasonable to conclude that with very high probability $|\hat{V}-V|/V$ is much smaller than .06. Analyses of errors in the substate allocations will be performed under the assumption that $|\hat{V}-V|/V$ is negligible. It should be kept in mind that the resulting estimates of errors in allocation will tend to underestimate the magnitude of these errors.

It will also be assumed that $\hat{O}_j-O_j$ is negligible. This assumption is minor because of the small size of the amounts $\hat{O}_j$ (calculations on EP 1 and EP 6 New Jersey data show $\hat{O}_j = 0$).

Under these assumptions the delta method will be applied to the substate allocation formulas presented in Table 5.2 to yield approximations for errors and moments of errors in substate allocations. The applicability of the delta method (or "modified" delta method) to piecewise smooth functions has already been discussed (§2.3). Furthermore, under the assumptions made above, each formula in Table 5.2 is a piecewise rational function of the data elements such that the denominator has non-negative coefficients and the function is bounded. It follows from the specification of the error models (chapter 3) and the discussion of §2.3 that assumptions (2.3)-(2.6) of the delta method are satisfied.

Because the quantities $<c/C>^*$ and c/C are close with high probability (see §3.15.4) we will proceed in the sequel as if the moments of $c_j/C_j - <c/C>^*$ and $c_j/C_j - c/C$ were the same. The reasoning is the same as that in §4.4 leading to (4.28).

In the sequel we will also proceed as if

$$\text{(5.15)} \qquad \mathrm{Cov}(c_{jk\alpha}/C_{jk\alpha} - <c/C>^{**}_{j\alpha}\ ,\ c_j/C_j - c/C)$$

was negligible. This is justified because (3.41) is small, and the difference between (3.41) and (5.15) is small, being a weighted average of small terms.

$$\sum_{K\alpha} w_{jk\alpha}\mathrm{Cov}(c_{jk\alpha}/C_{jk\alpha} - c_j/C_j\ ,\ c_j/C_j - c/C)$$

where $w_{jk\alpha}$ are non-negative and $\sum_{K\alpha} w_{jk\alpha} = 1$ (see (3.55)).

For discussion of the remainder terms in the approximations derived by the delta method, the reader is referred to §4.4 of the state-level analyses.

§5.5 Conclusions

Recall that $\langle d/D\rangle^*$, $\langle d/D\rangle^{**}_{j\alpha}$, $\langle c/C\rangle^*$, $\langle c/C\rangle^{**}_{j\alpha}$, ξ_j , $\xi_{jk\alpha}$, and $\phi_{j\alpha}$ were defined by (3.56)-(3.62) and that μ_Y and σ_Y^2 are the relative bias and variance of the allocation to New Jersey (see (4.5)-(4.6)).

Proposition 5.1

Under the assumptions of §5.4 and the definitions of V , X_j , $X_{jk\alpha}$, $\sigma_\xi^2(j)$, etc. made in §5.2, the following properties hold:

(i) the relative error in allocation to an unconstrained county area j is

(5.31) $$[\frac{y}{Y} + \xi_j] + o_p(N^{-1/2})$$

with mean

(5.32) $$\mu_Y - 2\mu_C(j) + o(N^{-1/2})$$

and variance

(5.33) $$\sigma_Y^2 + \sigma_\xi^2(j) + o(N^{-1})$$

where ξ_j and $\sigma_\xi^2(j)$ are defined in (3.60) and (5.7).

(ii) The relative error in allocation to PCLS-constrained county area j is

$$(5.34) \qquad \left(\frac{p_j}{P_j} - \frac{p.}{P.}\right) + \frac{y}{Y} + o_p(N^{-1/2})$$

with mean

$$(5.35) \qquad \mu_P(j) + \mu_Y + o(N^{-1/2})$$

and variance

$$(5.36) \qquad \sigma_P^2(j) + \sigma_Y^2 + o(N^{-1}) \quad .$$

(iii) The relative error in allocation to an unconstrained county government j in an unconstrained county area is

$$(5.37) \qquad (d_j/D_j - <d/d>*) - 2(c_j/C_j - <c/C>*) + \frac{y}{Y} + o_p(N^{-1/2})$$

with mean

$$(5.38) \qquad \mu_Y + o(N^{-1/2})$$

and variance

$$(5.39) \qquad \left(1 - \frac{w_j D_j}{D_j + D_j.}\right)^2 \sigma_D^2(j)$$

$$+ \sum_{\substack{\ell \epsilon J \\ \ell \neq j}} \left(\frac{w_\ell^2}{(D_\ell + D_\ell .)^2} \; [D_\ell^2 \, \sigma_D^2(\ell) + \sum_{k \epsilon K} D_{\ell k}^2 \, \sigma_D^2(\ell,k)]\right)$$

$$+ \; 4\,\sigma_C^2(j) + \sigma_Y^2 + O(N^{-1})$$

where $<d/D>*$, $<c/C>*$, and w_j are defined in (3.56), (3.58), and (3.54)

(iv) The relative error in allocation to county government j not constrained by a 50% ceiling but in a PCLS-constrained county area is

(5.40) $$\left(\frac{p_j}{P_j} - \frac{p.}{P.}\right) + \left(\frac{d_j}{D_j} - \frac{d_j+d_{j}.}{D_j+D_{j}.}\right) + \frac{y}{Y} + o_p(N^{-1/2})$$

with mean

(5.41) $$\mu_P(j) + \mu_Y + o(N^{-1/2})$$

and variance

(5.42) $$\sigma_P^2(j) + \sigma_{DD}^2(j) + \sigma_Y^2 + o(N^{-1}) \quad ,$$

where $\sigma_{DD}^2(j)$ is defined in (5.8).

(v) The relative error in allocation to county government j constrained by a 50% ceiling is

(5.43) $$\frac{d_j+g_j}{D_j+G_j} + o_p(N^{-1/2})$$

with mean

(5.44) $$o(N^{-1/2})$$

and variance

(5.45) $$\frac{D_j^2\ \sigma_D^2(j) + G_j^2\ \sigma_G^2(j)}{(D_j + G_j)^2} + o(N^{-1})$$

(vi) The relative error in allocation to unconstrained local government $jk\alpha$ in an unconstrained county area is

(5.46) $$\frac{y}{Y} + \xi_j + \xi_{jk\alpha} + \phi_{j\alpha} + o_p(N^{-1/2})$$

with mean

(5.47) $$\mu_Y - 2\mu_C(j) + \mu_\xi(j,k\alpha) + o(N^{-1/2})$$

and variance

(5.48) $$\sigma_Y^2 + \sigma_\xi^2(j) + \sigma_\xi^2(j,k\alpha) + \sigma_\phi^2(j,\alpha)$$
$$+ 2\,\sigma_{\xi\xi}(j,k\alpha) + 2\,\sigma_{\xi\phi}(j,\alpha) + 2\sigma'_{\xi\phi}(j,k\alpha) + o(N^{-1}) \quad ,$$

where ξ_j , $\xi_{jk\alpha}$, and $\phi_{j\alpha}$ are defined in (3.60), (3.61), and (3.62); $\mu_\xi(j,k\alpha)$ is defined in (5.14); and $\sigma_\xi^2(j)$, $\sigma_\xi^2(j,k\alpha)$, $\sigma_\phi^2(j,\alpha)$, $\sigma_{\xi\xi}(j,k\alpha)$, $\sigma_{\xi\phi}(j,k\alpha)$, and $\sigma'_{\xi\phi}(j,k\alpha)$ are defined in (5.7), (5.9), (5.10), (5.11), (5.12), and (5.13).

(vii) The relative error in allocation to an unconstrained local government $jk\alpha$ in a PCLS-constrained county area is

(5.49) $$\left(\frac{p_j}{P_j} - \frac{p.}{P.}\right) + \frac{y}{Y} + \xi_{jk\alpha} + \phi_{j\alpha} + o_p(N^{-1/2})$$

with mean

(5.50) $$\mu_P(j) + \mu_Y + \mu_\xi(j,k\alpha) + o(N^{-1/2})$$

and variance

$$(5.51) \qquad \sigma_P^2(j) + \sigma_Y^2 + \sigma_\xi^2(j,k\alpha) + \sigma_\phi^2(j,\alpha) + 2\sigma'_{\xi\phi}(j,k\alpha) + o(N^{-1})$$

where $\xi_{jk\alpha}$, $\phi_{j\alpha}$, and $\mu_\xi(j,k\alpha)$ are defined in (3.61), (3.62), and (5.14); and $\sigma_\xi^2(j,k\alpha)$, $\sigma_\phi^2(j,\alpha)$, and $\sigma'_{\xi\phi}(j,k\alpha)$ are defined in (5.9), (5.10), and (5.13).

(viii) The relative error in allocation to a PCLS-constrained local government $jk\alpha$ is

$$(5.52) \qquad \left(\frac{p_{jk\alpha}}{P_{jk\alpha}} - \frac{p.}{P.}\right) + \frac{y}{Y} + o_p(N^{-1/2})$$

with mean

$$(5.53) \qquad \mu_P(j,k\alpha) + \mu_Y + o(N^{-1/2})$$

and variance

$$(5.54) \qquad \sigma_P^2(j,k\alpha) + \sigma_Y^2 + o(N^{-1}) \quad .$$

(ix) The relative error in allocation to local government $jk\alpha$ constrained by a 50% ceiling is

$$(5.55) \qquad \frac{d_{jk\alpha}+g_{jk\alpha}}{D_{jk\alpha}+G_{jk\alpha}} + o_p(N^{-1/2})$$

with mean

$$(5.56) \qquad o(N^{-1/2})$$

and variance

$$\frac{D^2_{jk\alpha}\ \sigma^2_D(j,k\alpha) + G^2_{jk\alpha}\ \sigma^2_G(j,k\alpha)}{(D_{jk\alpha} + G_{jk\alpha})^2} + o(N^{-1}) \quad . \tag{5.57}$$

Proof:

The first steps in the proof consist of verifying that (5.7)-(5.14) are approximations to the variances of ξ_j , $\xi_{jk\alpha}$ etc. This verification is mostly straightforward algebra with exploitation of numerous absences of correlation; in particular it follows from discussion in §5.4 that

$$\mathrm{Cov}(c_j/C_j - \langle c/C\rangle^* \ , \ y/Y) = o(N^{-1})$$

and that

$$\mathrm{Cov}(c_{jk\alpha}/C_{jk\alpha} - \langle c/C\rangle^{**}_{j\alpha} \ , \ c_j/C_j - \langle c/C\rangle^*) = o(N^{-1}) \quad .$$

Also, use is made of the relations (from §5.4)

$$\mathrm{Var}(c_j/C_j - \langle c/C\rangle^*) = \sigma^2_C(j)$$

$$E(c_j/C_j - \langle c/C\rangle^*) = \mu_C(j) \quad .$$

Otherwise the proposition is proved much the same way as proposition 4.1.

*****

Notice that proposition 5.1 was proved under the assumption that the bias in $\hat{V}$ was negligible. In practice, by taking into account biases in the amounts redistributed (in the "inner" allocations) from PCLS-constrained to unconstrained county areas, the estimates of biases in allocations to unconstrained county areas and unconstrained units in unconstrained county areas (see (5.32), (5.38), (5.47)) can be adjusted for a large part of whatever bias is present in $\hat{V}$. Let Y_j denote the allocation to county area j and let $\mathcal{J}c$ ($\mathcal{J}u$) denote the set of PCLS-constrained (-unconstrained) county areas. Since (see §5.1) the additional amount ultimately distributed from constrained county areas to an unconstrained area j is proportional to Y_j, the bias in this amount may be expressed as βY_j. To approximate β note that the ratio $\hat{Y}_L/(2\hat{Y}/3)$ is insensitive to small changes in the data (for New Jersey it varied by less than 1.1% from EP's 1-6; see SRI 1974b, Table D-2).

From the identity

$$\frac{\Sigma\ \hat{Y}_j - Y_j}{\Sigma\ Y_j} = \frac{(\hat{Y}_L/(2\hat{Y}/3))\hat{Y} - (Y_L/(2Y/3))Y}{(Y_L/(2Y/3)Y}$$

$$= \frac{\hat{Y}-Y}{Y} - (1 - \frac{\hat{Y}_L/(2\hat{Y}/3)}{Y_L/(2Y/3)})\ \frac{\hat{Y}}{Y}$$

it now follows that $(\hat{Y}-Y)/Y$ is approximately equal to

$$\frac{\Sigma\ \hat{Y}_j - Y_j}{\Sigma\ Y_j}$$

and therefore the following approximate equality holds:

$$\mu_Y = \frac{E(\Sigma\ \hat{Y}_j - Y_j)}{\Sigma\ Y_j}\qquad .$$

Formulas (5.32) and (5.35) thus imply

$$\mu_Y = \frac{\sum_{Ju} Y_j(\mu_Y + \beta - 2\mu_C(j)) + \sum_{Jc} Y_j(\mu_Y + \mu_P(j))}{\sum Y_j}$$

and so

$$\beta = \frac{2\sum_{Ju} Y_j\mu_C(j) - \sum_{Jc} Y_j\mu_P(j)}{\sum_{Ju} Y_j} \quad .$$

Of course, β is smaller than the main relative bias terms; for EP 1 the estimated values of μ_Y and β are 0.0254 and 0.0022. This correction for β will be used in estimating moments of errors in substate allocations (chapter 6).

Chapter 6 Computations and Analyses

§6.1 Introduction

We are now ready to capitalize on the laborious analyses of earlier chapters. We begin by determining the sizes of the biases and standard deviations of the GRS fund allocations to state areas (§6.2) and various substate jurisdictions in New Jersey (§6.3). These estimates derive from the formulas developed in chapters 4 and 5 upon substitution of the estimates developed in chapter 3 for the biases and variances of the data elements.

By considering hypothetical alternative error models for the data we also study the sensitivity of the accuracy of the allocations to changes in the quality of data. We will analyze the consequences of improving several data series. Probably of greatest interest are the implications of reducing population undercount. In particular, we measure what the effects on GRS allocations would have been if the 1970 census undercoverage rates had been less severe or, alternatively, more severe. Similar sensitivity analyses are also performed for other data elements. The findings are summarized in §6.5 below.

The approach taken is to vary the quality of the data elements, one or two at a time, and observe the effects on moments of misallocations. We adopt this naive approach instead of using more sophisticated experimental designs because the dependences of the moments of misallocation upon the moments of the data elements are complex. Also, the estimated relationships are individually interesting.

Several benefit analyses are then performed (§ 6.4) to compare the benefits and costs of data improvement programs. One analysis considers whether more money should have been spent to reduce undercount in the 1970 census. Other analyses focus on: improving money income data; improving the accuracy of projections of state individual income tax collections; using revised rather than provisional estimates of state populations in entitlement periods 3 and 4; and improving the accuracy of the estimates of adjusted taxes for substate governments in New Jersey. The findings are briefly summarized (§6.5) and their implications for policy are discussed in chapter 7 below.

The reader uninterested in errors in General Revenue Sharing allocations per se is encouraged to skip §§6.2 and 6.3 and to proceed directly to §6.4.

To measure the benefit of more accurate allocations arising from improved data quality we study the sum of expected absolute misallocations. Rationale for using this measure was discussed in chapter 1, especially in §1.6 and at the end of §1.7. Because our concern here is with economics of misallocations (in contrast to a statistically motivated error analysis) the emphasis will be on dollar misallocations rather than relative errors in allocation. Thus the tables in this chapter present estimates of "pure" bias and standard deviation rather than relative bias or relative standard deviation.

Estimates of the sum of expected absolute errors in GRS allocations will be based on a total pie of $55 billion, spread over 11 entitlement periods (EP's). To see where this total comes from, note that in 1976 Public Law 94-488 extended the original GRS program to encompass four additional EP's. The same formulas as in EP's 1-7 determine the shares of the allocated pie. However, the pie is no longer fixed, but equals, on a twelve month basis, the minimum of $6.85 billion and $6.65 billion times the ratio of (i) federal individual income taxes collected in the last calendar year ending more than one year before the termination of the EP, to (ii) the amount of federal individual income taxes collected in calendar year 1975. The sizes of the pies in EP's 10,11 are not yet determined but it is reasonable to assume they will be no smaller than in EP 9. The following table is adapted from Office of Revenue Sharing (1977, p.vii).

Entitlement Period	Start	End	Duration	Amount (billions of dollars
1	Jan. 72	June 72	6 months	2.65
2	July 72	Dec. 72	6 months	2.65
3	Jan. 73	June 73	6 months	2.9875
4	July 73	June 74	1 year	6.05
5	July 74	June 75	1 year	6.2
6	July 75	June 76	1 year	6.35
7	July 76	Dec. 76	6 months	3.325
8	Jan. 77	Sept. 77	9 months	4.987
9	Oct. 77	Sept. 78	1 year	6.85
10	Oct. 78	Sept. 79	1 year	6.85*
11	Oct. 79	Sept. 80	1 year	6.85*
			TOTAL	55.73

Table 6.0

Total Allocations for GRS, EP's 1-11

* estimated

§6.2 Errors in Allocations to States

Moments of errors in allocations are estimated under several alternative models, including

Model

STANDARD	The error model derived in chapter 3. This model reflects the actual error structures in the data.
POP(λ)	Same as STANDARD except that all population undercoverage rates are multiplied by λ . For EP 1, this is equivalent to multiplying $\mu_P(i;1)$, $\mu_U(i;1)$, $\mu_P(i,j;1)$, and $\mu_P(i,j,k;1)$ by λ (see §3.2-§3.4). Note that POP(1) is the same as STANDARD.
MON(λ)	Same as STANDARD except all biases in estimates of state money income have been multiplied by λ . For EP 1, this is equivalent to multiplying $\mu_M(i;1)$ by λ (see §3.5). Note that MON(1) is the same as STANDARD.
POPMON(λ)	Same as STANDARD except all biases in estimates of state money income (§3.5) and all population undercoverage rates (§3.2-§3.4) are multiplied by λ . For EP 1, this is equivalent to multiplying $\mu_M(i;1)$, $\mu_M(i,j;1)$, $\mu_M(i,j,k;1)$, $\mu_P(i;1)$, $\mu_P(i,j;1)$, $\mu_P(i,j,k;1)$ and $\mu_U(i;1)$ by λ (see §3.2-§3.6). Note that POPMON(1) is the same as STANDARD.

STINCTX	Same as STANDARD except biases and variances of the EP 1, EP 2 estimates of state individual income tax collections (§3.9) have been set to zero. That is, $\mu_K(i;t) = \sigma_K^2(i;t) = 0$ for $t = 1,2$.
POP-UP	Same as STANDARD except that population updating variances (§3.2, §3.4) have been set to zero.
POP-REV	Same as STANDARD except that population revision variance (§3.2) has been set to zero.
POP-REVUP	Same as STANDARD except that population revision variance and population updating variance (§3.2, §3.4) have both been set to zero.
PCI-UP	Same as POP-UP except that the variance of the per capita income updates (§3.5 and §3.6) has also been set to zero.
ADJTAX(λ)	Same as STANDARD except that the variances of the relative errors in adjusted taxes have all been multiplied by λ. That is, $\sigma_D^2(i,j;t)$ and $\sigma_D^2(i,j,k;t)$ have been multiplied by λ (see §3.12).

Calculations were performed by FORTRAN programs on Yale University's IBM 370/158 computer. Values of state-level data elements were obtained from Office of Revenue Sharing (1973-1976) while values of substate data elements were taken from the tape "General Revenue Sharing-Data Elements" available from the Data User Services Division of the U.S. Bureau of the Census. With a single exception, the programming to calculate the estimates of means, standard deviations, and expected absolute values of errors in allocations (from the formulas in chapters 4 and 5) was straightforward. The exception was the calculation of the substate allocations, for which the FORTRAN program "Interstate Allocation Program, Version 4.45 - IBM 360/370" (available from Westat, Inc.) was used.

Biases, variances, and expected absolute errors in state allocations were computed under a variety of models.

EP	Models
1	POP(λ) , MON(λ) , POPMON(λ) STINCTX , STANDARD
3	POP-UP , POP-REVUP , POP-REV , STANDARD
4	POP-UP , POP-REVUP , POP-REV , STANDARD
6	POP-UP , PCI-UP , STANDARD

Table 6.1

Error Models Used for Sensitivity Analysis of State Allocations

Moments of errors were not estimated for EP's 2,5 or 7 because of the similarity of the stochastic structures underlying the respective data elements to those for the immediately preceding EP.

Formula (3.8) implies that the effects of population undercoverage on biases in allocations are rather constant for EP's 1-7, since for each state the ratio of EP 1 population size to EP t population size is nearly constant for $t \leq 7$. Thus biases in EP 6 allocations will react comparably to biases in EP 1 allocations when the undercoverage rates are altered (Models POP(λ) $0 \leq \lambda \leq 2$).

Similarly (see formula (3.22)), biases in EP 6 allocations will react comparably to biases in EP 1 allocations when biases in EP 1 money income estimates are altered (Models MON(λ) $0 \leq \lambda \leq 2$) and when biases in EP 1 estimates of both money income and population are altered (Models POPMON (λ) $0 \leq \lambda \leq 2$).

Also, recall from §3.2 and §3.5 that in EP's 1 and 2 population and per capita income estimates were not updated and no revision error was present, so Models POP-UP , POP-REV , POP-REVUP , PCI-UP are not relevant for those EP's. In EP 6 both population and per capita income were updated but again no revision error was present, so Models POP-REVUP and POP-UP are equivalent in EP 6 as are Models POP-REV and STANDARD.

Tables 6.2 and 6.3 give estimates of moments of the misallocations for each state in EP's 1 and 6 under some of the error models described above. State by state estimates are not displayed for all the error models, but Tables 6.4 and 6.5 do present estimates of the sum of the expected absolute misallocations to states under all the error models.

In Tables 6.2 and 6.3 states with large expected gains (i.e. positive biases) under the STANDARD model include Illinois, Iowa, Kansas, Massachusetts, Minnesota, Nebraska, New York, Ohio, Pennsylvania and Wisconsin, while those with large expected losses include Arizona, California, Florida, Georgia, New Mexico, South Carolina, Tennessee and Texas. For Nebraska and Florida the absolute biases exceeded 10% of the EP 1 allocation. The figures in these tables are slightly larger than, but on the whole consistent with, those obtained by Siegel (1975b), who used estimates of undercoverage that varied comparatively little over states.

If population undercoverage were the sole source of bias in the allocations then the bias in a state's allocation would vary monotonically with λ and would not change sign for $\lambda > 0$. Thus the absolute bias in each state's allocation would be an increasing function of λ . In fact, population undercoverage is not the sole source of bias, and when the net bias from non-undercoverage sources has opposite sign the above patterns will not occur.

Table 6.2 shows that the bias in allocation to 12 states changed sign as λ varied from 0 to 1 (AR, CA, DC, HA, IN, MD, MT, NV, NM, VT, WI, WY) and the bias in allocation to 3 states changed sign as λ varied from 1 to 2 (NC, UT, WA). For eight states the absolute bias in allocation decreased monotonically for $0 \leq \lambda \leq 2$ (AL, CO, DE, LA, ME, MO, NH, OK). If λ were increased beyond 2 the biases in allocation to each of these eight states would eventually change sign as population undercoverage became the dominant bias. Thus improvement in the undercoverage rates will not reduce the absolute bias in allocation to all states. It will be seen from Table 6.4 that the overall inequity is improved when the undercoverage rates are improved (i.e. when λ is decreased in $POP(\lambda)$).

The standard deviations of the allocations behave more consistently, and in Table 6.3 the ratio of columns 4 to 5 and columns 5 to 6 are roughly 1.02 and 1.3. This consistency follows from the lack of correlation of population updating errors and per capita income updating errors with other errors in data, so that no "offsetting" of variances occurs (in contrast to the offsetting of biases exhibited in Table 6.2). Reducing these components of variance (the updating errors) therefore reduces the standard deviation (and the expected absolute deviation) in allocation to all states.

State	Allocation	Bias				Standard Deviation	
		POP(0)	STANDARD	POP(2)	STINCTX	STANDARD	STINCTX
AL	45.27	-1.38	-0.79	-0.21	-0.80	1.57	1.55
AK	3.30	-0.07	-0.26	-0.46	-0.26	0.12	0.11
AZ	25.11	-1.01	-2.20	-3.38	-2.20	0.88	0.87
AR	27.26	0.81	-0.31	-1.44	-0.32	0.95	0.94
CA	280.08	0.69	-2.88	-6.42	-2.54	5.31	2.68
CO	27.26	-0.28	-0.16	-0.03	-0.13	0.60	0.23
CT	33.60	0.46	1.09	1.73	0.87	0.35	0.33
DE	8.03	0.06	0.04	0.01	0.05	0.29	0.06
DC	11.95	0.16	-0.03	-0.23	-0.01	0.39	0.09
FL	73.32	-5.16	-9.50	-13.84	-9.51	2.53	2.50
GA	54.75	-0.02	-2.49	-4.97	-2.51	1.90	1.88
HA	11.85	0.04	-0.17	-0.36	-0.14	0.42	0.12
ID	10.64	-0.43	-0.57	-0.71	-0.57	0.37	0.37
IL	136.98	1.09	2.07	3.02	2.22	2.68	1.15
IN	56.87	-0.02	1.50	3.03	1.49	1.97	1.95
IO	37.73	1.49	2.81	4.12	2.80	1.31	1.30
KA	26.21	1.04	2.06	3.07	2.05	0.91	0.90
KY	43.47	-0.45	-0.50	-0.55	-0.51	1.51	1.49
LA	61.24	1.81	1.62	1.42	1.61	2.12	2.10
ME	15.99	0.31	0.30	0.28	0.29	0.56	0.55
MD	53.54	-0.08	0.27	0.59	0.37	1.57	0.45
MA	82.55	0.43	1.79	3.18	1.94	2.28	0.71
MI	112.20	0.24	0.86	1.50	1.02	2.57	0.95
MN	53.21	1.07	2.17	3.28	1.19	0.61	0.48
MS	44.21	1.31	1.88	2.44	1.87	1.53	1.52
MO	49.16	-0.14	-0.10	-0.08	-0.06	0.80	0.33
MT	10.24	-0.11	0.25	0.61	0.25	0.36	0.35
NB	19.43	1.35	2.07	2.78	2.06	0.68	0.67
NV	5.76	0.07	-0.10	-0.29	-0.13	0.07	0.07
NH	8.29	-0.34	-0.23	-0.12	-0.23	0.29	0.29
NJ	83.31	0.76	2.12	3.46	1.65	0.79	0.72
NM	16.48	0.16	-1.71	-3.57	-1.71	0.58	0.57
NY	294.61	2.20	4.20	6.21	4.74	6.89	2.91
NC	67.98	0.65	0.03	-0.59	0.02	2.35	2.32
ND	11.08	0.11	0.51	0.92	0.51	0.39	0.38
OH	106.96	0.29	1.78	3.28	1.83	1.28	0.69
OK	29.45	-1.48	-0.81	-0.14	-0.82	1.03	1.02
OR	26.50	0.04	0.31	0.58	0.36	0.87	0.20
PA	138.96	0.61	2.89	5.17	3.05	2.74	1.07
RI	12.08	0.03	0.23	0.44	0.25	0.24	0.10
SC	36.04	-0.01	-2.07	-4.14	-2.08	1.25	1.24
SD	12.06	0.72	0.80	0.88	0.80	0.42	0.42
TN	49.40	-1.01	-2.05	-3.10	-2.06	1.71	1.70
TX	123.92	-1.29	-6.88	-12.48	-6.91	4.21	4.17
UT	15.28	-1.08	-0.24	0.60	-0.24	0.53	0.53
VT	7.35	-0.00	0.11	0.22	0.10	0.26	0.26
VA	53.16	-0.40	-0.45	-0.49	-0.37	1.28	0.35
WA	38.98	-0.80	-0.37	0.05	-0.38	1.35	1.34
WV	25.96	-0.27	-1.29	-2.30	-1.29	0.90	0.90
WI	66.60	-2.03	2.29	6.61	2.28	2.30	2.28
WY	4.98	-0.15	0.08	0.32	0.08	0.17	0.17

Table 6.2

Moments of Errors in GRS Allocations to States, EP1

(all figures in millions of dollars)

State	Allocation	Bias	Standard Deviation		
		STANDARD	STANDARD	POP-UP	PCI-UP
AL	101.57	-1.50	4.61	4.51	3.49
AK	9.18	-0.14	0.09	0.08	0.06
AZ	64.04	-4.50	2.91	2.85	2.21
AR	66.13	-1.30	3.01	2.94	2.28
CA	659.77	-6.98	7.67	7.34	6.45
CO	69.03	-0.19	0.77	0.71	0.54
CT	85.46	2.06	1.13	1.07	0.94
DE	19.24	0.08	0.19	0.18	0.14
DC	26.79	-0.17	0.25	0.23	0.20
FL	196.32	-19.87	8.83	8.64	6.70
GA	131.82	-6.23	5.97	5.83	4.52
HA	27.82	-0.38	0.32	0.30	0.26
ID	24.49	-1.07	1.12	1.09	0.85
IL	323.23	4.44	3.50	3.26	2.63
IN	128.67	1.93	1.51	1.37	0.93
IO	84.41	5.64	3.84	3.75	2.91
KA	58.31	3.91	2.65	2.60	2.01
KY	103.85	-1.23	4.71	4.61	3.57
LA	136.97	2.49	6.20	6.06	4.70
ME	39.98	0.58	1.82	1.78	1.38
MD	126.30	0.72	1.36	1.28	1.06
MA	206.54	4.29	2.33	2.22	1.93
MI	267.44	2.09	3.06	2.87	2.34
MN	133.00	5.67	6.02	5.89	4.56
MS	93.69	2.69	4.25	4.16	3.23
MO	122.29	-1.97	5.54	5.42	4.20
MT	23.51	0.51	1.07	1.05	0.81
NB	42.38	3.74	1.93	1.89	1.47
NV	14.96	-0.78	0.68	0.67	0.52
NH	20.42	-0.42	0.93	0.91	0.71
NJ	199.35	3.59	2.29	2.14	1.77
NM	39.64	-3.89	1.81	1.77	1.37
NY	720.73	9.90	8.56	8.33	7.66
NC	155.16	-1.03	7.01	6.85	5.31
ND	20.48	0.85	0.94	0.91	0.71
OH	259.46	4.11	2.96	2.67	1.81
OK	69.46	-1.34	3.16	3.09	2.40
OR	66.67	0.78	0.72	0.67	0.51
PA	338.79	14.32	15.00	14.70	11.40
RI	27.53	0.54	0.31	0.28	0.22
SC	88.78	-5.13	4.03	3.94	3.06
SD	25.15	1.31	1.15	1.12	0.87
TN	118.87	-4.61	5.39	5.27	4.08
TX	305.72	-15.67	13.60	13.30	10.30
UT	37.61	-0.18	1.72	1.68	1.30
VT	18.57	0.24	0.85	0.83	0.64
VA	127.98	-0.81	1.40	1.27	0.88
WA	93.83	-0.42	4.26	4.17	3.23
WV	57.53	-2.87	2.62	2.56	1.99
WI	161.00	6.07	7.27	7.11	5.51
WY	10.08	0.16	0.46	0.45	0.35

Table 6.3

Moments of Errors in GRS Allocations to States, EP6
(all figures in millions of dollars)

EP 1. Total Allocation (in millions of dollars) = 2,650

Error Model	Sum of Mean Absolute Errors (millions of dollars)	Sum of Mean Absolute Errors Divided by $2,650 million (×100)
POP(0)	67.8	2.558
MON(0)	77.5	2.924
POPMON(0)	55.8	2.106
POP(.5)	75.9	2.864
MON(.5)	82.7	3.122
POPMON(.5)	67.0	2.528
STINCTX	80.6	3.041
STANDARD	91.1	3.438
POP(1.5)	110.6	4.175
MON(1.5)	102.0	3.849
POPMON(1.5)	120.7	4.554
POP(2.0)	132.5	5.001
MON(2.0)	114.7	4.329
POPMON(2.0)	152.9	5.771

Table 6.4

Sums of Mean Absolute Errors in GRS Allocations to States, EP 1

Error Model	Sum of Mean Absolute Errors (millions of dollars)	Sum of Mean Errors as % of Total Total EP Allocation
EP 3		
STANDARD	93.6	3.134
POP-UP	93.1	3.116
POP-REV	93.3	3.124
POP-REVUP	92.8	3.106
EP 4		
STANDARD	190.9	3.156
POP-UP	189.8	3.138
POP-REV	190.3	3.146
EP 6		
STANDARD	208.8	3.288
POP-UP	206.4	3.250
PCI-UP	190.2	2.995

EP 3 allocation = $2,987.5 million

EP 4 allocation = $6,050 million

EP 6 allocation = $6,350 million

Table 6.5

Sums of Mean Absolute Errors in GRS Allocations to States, EP's 3,4,6

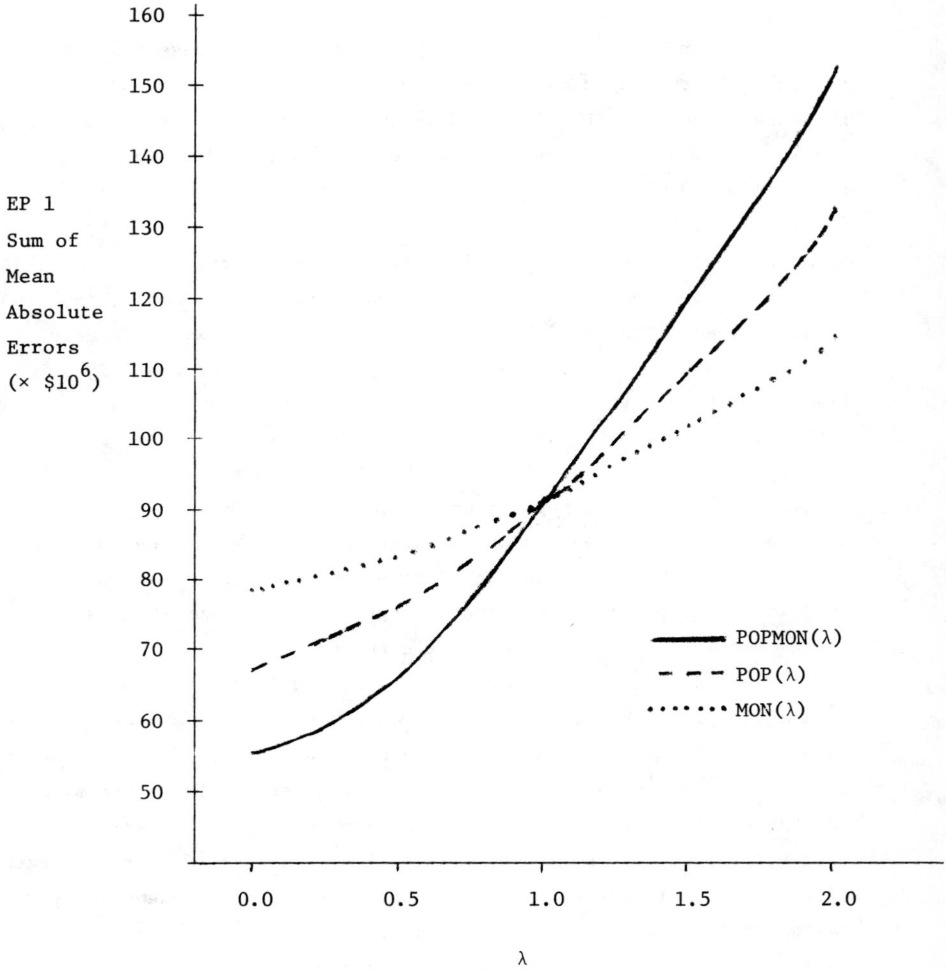

Figure 6.1

(Based on data in Table 6.4 and the assumption that the curves are convex)

Note from Table 6.4 that if population undercoverage could be eliminated, the expected sum of absolute misallocations to state areas would be reduced in EP 1 by \$23.3 million. Apparently, the sum of expected absolute misallocations under the error models POP(λ) is a convex function of $0 < \lambda < 2$, which would imply that for $0 < \lambda < 1$ (or $1 < \lambda < 2$) multiplying all undercoverage rates uniformly by λ has the effect of reducing (or increasing) the expected sum of misallocations by more than $|1-\lambda|\cdot$\$23.3 million. Thus, for example, the reduction obtained when $\lambda=.5$ is \$15.2(= 91.1-75.9) million, significantly greater than half of \$23.3(= 91.1-67.8) million. The graph in Figure 6.1 illustrates that the expected sum of absolute misallocations varies convexly in λ for MON(λ) and POPMON(λ) as well as for POP(λ).

Figure 6.1 (or equivalently, Table 6.4) indicates that of reducing the biases in population data, money income data or both, the greatest improvement in the accuracy of the allocations comes from reducing the biases in both data elements, with population next in importance and money income last. Of course these data improvement programs would have different costs.

It is interesting to compare these conclusions with those of Siegel (1975b, p.20), who considered the effects of biases in population data, per capita income data, and both. The difference in emphasis between Siegel's (population and per capita income) and the present study (population and total money income) arises from a difference in perspective; Siegel is concerned with the effects of adjusting existing estimates for biases while the present study is concerned with the effects of reducing biases in data not yet collected. For although population biases are caused by persons <u>not</u> enumerated, the biases in money income data arise predominantly from the underreporting of income of people who <u>were</u> enumerated. Thus the major biases in the two data elements arise separately and it is meaningful to consider their reduction separately.

Siegel's results are reproduced as Tables 6.6 and 6.7. For the first table the synthetic method was employed to estimate the undercount according to age, race, and sex as given in Census (1974a). The errors in per capita income were estimated by considering underreporting of money income, in a manner similar to that in the present study (§3.5.4). The table presumes a fixed pie of $5.3 billion. The second table is based on similar procedures except that the synthetic method was modified to account for the inverse relationship between undercount and median family income (see Siegel 1975b, pp.10-11 for details).

Of existing biases in population data, per capita income data and both, Siegel found that adjusting the biases in both jointly to have the greatest effect on allocations, with per capita income next in importance, followed by population. Siegel's findings and those of the present study are not comparable in an exact numeric sense partly because different error models were used. Also, Siegel did not look at deviations of allocations from paramenter values (as in the present study) but focused instead on the deviations of allocations based on adjusted (for biases) data from allocations based on unadjusted data. However, the sum of the absolute deviations (in Siegel's sense) arising from adjusted data for biases can be roughly interpreted as the reduction in expected sum of absolute misallocations when biases in the data have been altered.

Taken together, Siegel's and the present study indicate that changes in biases in population and income jointly are the most significant, followed in order of decreasing importance by per capita income, population, and money income. The greater significance of per capita income biases than either population or money income biases can be attributed to the dominance of underreporting over undercoverage as the major contributor to bias in money income estimates. (For more discussion see Technical Appendix).

Region, division, and State	Census[1]	Corrected population only[2]	Corrected per capita income	Corrected population[2] and per capita income	Difference from census: Corrected population only — Amount	Percent	Corrected per capita income — Amount	Percent	Corrected population and per capita income — Amount	Percent
United States, total	5,300,000	5,300,000	5,300,000	5,300,000	-	-	-	-	-	-
Northeast	1,362,034	1,360,954	1,354,551	1,353,474	-1,080	-0.1	-7,483	-0.5	-8,560	-0.6
New England	322,209	321,108	320,991	319,896	-1,101	-0.3	-1,218	-0.4	-2,313	-0.7
Maine	32,678	32,458	31,382	31,171	-220	-0.7	-1,296	-4.0	-1,507	-4.6
New Hampshire	16,544	16,437	17,017	16,908	-107	-0.6	+473	+2.9	+364	+2.2
Vermont	14,855	14,759	14,875	14,779	-96	-0.6	+20	+0.1	-76	-0.5
Massachusetts	166,005	165,535	165,447	164,978	-470	-0.3	-558	-0.3	-1,027	-0.6
Rhode Island	24,255	24,168	24,150	24,062	-87	-0.4	-105	-0.4	-193	-0.8
Connecticut	67,872	67,751	68,120	67,998	-121	-0.2	+248	+0.4	+126	+0.2
Middle Atlantic	1,039,825	1,039,846	1,033,560	1,033,578	+21	-	-6,265	-0.6	-6,247	-0.6
New York	593,420	593,603	587,520	587,698	+183	-	-5,900	-1.0	-5,722	-1.0
New Jersey	167,380	167,370	168,398	168,388	-10	-	+1,018	+0.6	-1,008	+0.6
Pennsylvania	279,025	278,873	277,642	277,492	-152	-0.1	-1,383	-0.5	-1,533	-0.5
North Central	1,378,138	1,376,007	1,366,908	1,364,835	-2,131	-0.2	-11,230	-0.8	-13,303	-1.0
East North Central	962,671	962,144	961,000	960,463	-527	-0.1	-1,671	-0.2	-2,208	-0.2
Ohio	214,929	214,879	214,239	214,189	-50	-	-690	-0.3	-740	-0.3
Indiana	114,321	114,035	114,185	113,901	-286	-0.3	-136	-0.1	-420	-0.4
Illinois	275,147	275,483	272,917	273,250	+336	+0.1	-2,230	-0.8	-1,897	-0.7
Michigan	223,954	224,109	223,536	223,690	+155	+0.1	-418	-0.2	-264	-0.1
Wisconsin	134,320	133,638	136,123	135,433	-682	-0.5	+1,803	+1.3	+1,113	+0.8
West North Central	415,467	413,863	405,908	404,372	-1,604	-0.4	-9,559	-2.3	-11,095	-2.7
Minnesota	106,828	106,459	106,579	106,211	-369	-0.3	-249	-0.2	-617	-0.6
Iowa	76,508	76,045	73,413	72,970	-463	-0.6	-3,095	-4.0	-3,538	-4.6
Missouri	98,813	98,750	98,425	98,374	-63	-0.1	-388	-0.4	-439	-0.4
North Dakota	19,891	19,752	19,470	19,334	-139	-0.7	-421	-2.1	-557	-2.8
South Dakota	23,633	23,456	21,878	21,715	-177	-0.7	-1,755	-7.4	-1,918	-8.1
Nebraska	38,389	38,195	35,595	35,415	-194	-0.5	-2,794	-7.3	-2,974	-7.7
Kansas	51,405	51,206	50,548	50,353	-199	-0.4	-857	-1.7	-1,052	-2.0
South	1,605,758	1,612,893	1,619,428	1,626,540	+7,135	+0.4	+13,670	+0.9	+20,782	+1.3
South Atlantic	762,690	766,518	775,934	779,790	+3,828	+0.5	+13,244	+1.7	+17,100	+2.2
Delaware	16,066	16,082	15,824	15,839	+16	+0.1	-242	-1.5	-227	-1.4
Maryland	107,297	107,596	108,068	108,370	+299	+0.3	+771	+0.7	+1,073	+1.0
District of Columbia	23,937	24,465	23,551	24,065	+528	+2.2	-386	-1.6	+128	+0.5
Virginia	106,383	106,701	107,427	107,751	+318	+0.3	+1,044	+1.0	+1,368	+1.3
West Virginia	50,903	50,630	51,167	50,893	-273	-0.5	+264	+0.5	-10	-
North Carolina	135,591	136,444	133,724	134,567	-853	+0.6	-1,867	-1.4	-1,024	-0.8
South Carolina	71,580	72,383	71,801	72,608	+803	+1.1	+221	+0.3	+1,028	+1.4
Georgia	108,049	109,023	109,431	110,418	+974	+0.9	+1,382	+1.3	+2,369	+2.2
Florida	142,884	143,194	154,941	155,279	+310	+0.2	+12,057	+8.4	+12,395	+8.7
East South Central	359,764	361,619	360,972	362,813	+1,855	+0.5	+1,208	+0.3	+3,049	+0.8
Kentucky	85,722	85,492	85,723	85,494	-230	-0.3	+1	-	-228	-0.3
Tennessee	98,049	98,278	99,881	100,115	+229	+0.2	+1,832	+1.9	+2,066	+2.1
Alabama	89,225	89,932	91,806	92,534	+707	+0.8	+2,581	+2.9	+3,309	+3.7
Mississippi	86,768	87,917	83,562	84,670	+1,149	+1.3	-3,206	-3.7	-2,098	-2.4
West South Central	483,304	484,756	482,522	483,937	+1,452	+0.3	-782	-0.2	+633	+0.1
Arkansas	52,670	52,827	50,790	50,942	+157	+0.3	-1,880	-3.6	-1,728	-3.3
Louisiana	122,477	123,769	119,512	120,774	+1,292	+1.1	-2,965	-2.4	-1,703	-1.4
Oklahoma	59,025	58,829	61,226	61,024	-196	-0.3	+2,201	+3.7	+1,999	+3.4
Texas	249,132	249,331	250,994	251,197	+199	+0.1	+1,862	+0.7	+2,065	+0.8
West	954,071	950,143	959,115	955,153	-3,928	-0.4	+5,044	+0.5	+1,082	+0.1
Mountain	227,079	225,791	231,729	230,414	-1,288	-0.6	+4,650	+2.0	+3,335	+1.5
Montana	20,167	20,025	20,030	19,889	-142	-0.7	-137	-0.7	-278	-1.4
Idaho	21,118	20,972	21,467	21,320	-146	-0.7	+349	+1.7	+202	+1.0
Wyoming	9,542	9,478	9,539	9,475	-64	-0.7	-3	-	-67	-0.7
Colorado	54,307	54,126	54,862	54,678	-181	-0.3	+555	+1.0	+371	+0.7
New Mexico	32,817	32,592	32,460	32,238	-225	-0.7	-357	-1.1	-579	-1.8
Arizona	47,707	47,410	49,868	49,558	-297	-0.6	+2,161	+4.5	+1,851	+3.9
Utah	29,907	29,698	32,024	31,801	-209	-0.7	+2,117	+7.1	+1,894	+6.3
Nevada	11,514	11,490	11,479	11,455	-24	-0.2	-35	-0.3	-59	-0.5
Pacific	726,992	724,352	727,386	724,739	-2,640	-0.4	+394	+0.1	-2,253	-0.3
Washington	80,785	80,333	81,607	81,151	-452	-0.6	+822	+1.0	+366	+0.5
Oregon	53,061	52,859	53,510	53,304	-202	-0.4	+449	+0.8	+243	+0.5
California	563,150	561,638	562,432	560,920	-1,512	-0.3	-718	-0.1	-2,230	-0.4
Alaska	6,227	6,211	6,195	6,178	-16	-0.3	-32	-0.5	-49	-0.8
Hawaii	23,769	23,311	23,642	23,186	-458	-1.9	-127	-0.5	-583	-2.5

- Zero or rounds to less than ± 0.05 percent.

[1]Distribution differs from official entitlements published by the Office of Revenue Sharing partly because the official entitlements exclude money held in reserve.

[2]Pertains to the general population factor and the urbanized population factor.

Table 6.6

(adopted from Siegel(1975b,p 27); numbers are in thousands)

GRS Allocations to States Based on 1970 Census Reported Data and 1970 Census Data Corrected by the Basic Synthetic Method

Region, division, and State	Census[1]	Corrected population only[2]	Corrected per capita income	Corrected population[2] and per capita income	Difference from census					
					Corrected population only		Corrected per capita income		Corrected population and per capita income	
					Amount	Percent	Amount	Percent	Amount	Percent
United States, total	5,300,000	5,300,000	5,300,000	5,300,000	-	-	-	-	-	-
Northeast	1,362,034	1,356,962	1,351,285	1,346,219	-5,072	-0.4	-10,749	-0.8	-15,815	-1.2
New England	322,209	320,357	320,302	318,443	-1,852	-0.6	-1,907	-0.6	-3,766	-1.2
Maine	32,678	32,504	31,406	31,236	-174	-0.5	-1,272	-3.9	-1,442	-4.4
New Hampshire	16,544	16,399	16,967	16,816	-145	-0.9	+423	+2.6	+272	+1.6
Vermont	14,855	14,749	14,856	14,749	-106	-0.7	+1	-	-106	-0.7
Massachusetts	166,005	165,117	165,054	164,167	-888	-0.5	-951	-0.6	-1,838	-1.1
Rhode Island	24,255	24,133	24,100	23,977	-122	-0.5	-155	-0.6	-278	-1.1
Connecticut	67,872	67,455	67,919	67,498	-417	-0.6	+47	+0.1	-374	-0.6
Middle Atlantic	1,039,825	1,036,605	1,030,983	1,027,776	-3,220	-0.3	-8,842	-0.9	-12,049	-1.2
New York	593,420	591,797	586,076	584,476	-1,623	-0.3	-7,344	-1.2	-8,944	-1.5
New Jersey	167,380	166,557	167,889	167,056	-823	-0.5	+509	+0.3	-324	-0.2
Pennsylvania	279,025	278,251	277,018	276,244	-774	-0.3	-2,007	-0.7	-2,781	-1.0
North Central	1,378,138	1,371,546	1,363,141	1,356,614	-6,592	-0.5	-14,997	-1.1	-21,524	-1.6
East North Central	962,671	958,024	957,793	953,128	-4,647	-0.5	-4,878	-0.5	-9,543	-1.0
Ohio	214,929	214,021	213,626	212,717	-908	-0.4	-1,303	-0.6	-2,212	-1.0
Indiana	114,321	113,539	113,618	112,832	-782	-0.7	-703	-0.6	-1,489	-1.3
Illinois	275,147	274,275	272,119	271,256	-872	-0.3	-3,028	-1.1	-3,891	-1.4
Michigan	223,954	223,010	222,859	221,914	-944	-0.4	-1,095	-0.5	-2,040	-0.9
Wisconsin	134,320	133,179	135,571	134,409	-1,141	-0.8	+1,251	+0.9	+89	+0.1
West North Central	415,467	413,522	405,348	403,486	-1,945	-0.5	-10,119	-2.4	-11,981	-2.9
Minnesota	106,828	106,260	106,350	105,782	-568	-0.5	-478	-0.4	-1,046	-1.0
Iowa	76,508	75,978	73,302	72,788	-530	-0.7	-3,206	-4.2	-3,720	-4.9
Missouri	98,813	98,585	98,199	98,019	-228	-0.2	-614	-0.6	-794	-0.8
North Dakota	19,891	19,796	19,502	19,408	-95	-0.5	-389	-2.0	-483	-2.4
South Dakota	23,633	23,523	21,927	21,823	-110	-0.5	-1,706	-7.2	-1,810	-7.7
Nebraska	38,389	38,193	35,571	35,387	-196	-0.5	-2,818	-7.3	-3,002	-7.8
Kansas	51,405	51,187	50,497	50,279	-218	-0.4	-908	-1.8	-1,126	-2.2
South	1,605,758	1,622,996	1,628,795	1,646,028	+17,238	+1.1	+23,037	+1.4	+40,270	+2.5
South Atlantic	762,690	768,225	777,646	783,216	+5,535	+0.7	+14,956	+2.0	+20,526	+2.7
Delaware	16,066	16,037	15,786	15,758	-29	-0.2	-280	-1.7	-308	-1.9
Maryland	107,297	107,060	107,731	107,487	-237	-0.2	+434	+0.4	+190	+0.2
District of Columbia	23,937	24,155	23,443	23,654	+218	+0.9	-494	-2.1	-283	-1.2
Virginia	106,383	106,585	107,219	107,420	+202	+0.2	+836	+0.8	+1,037	+1.0
West Virginia	50,903	50,898	51,405	51,396	-5	-	+502	+1.0	+493	+1.0
North Carolina	135,591	137,193	134,374	135,950	+1,602	+1.2	-1,217	-0.9	+359	+0.3
South Carolina	71,580	73,097	72,463	73,992	+1,517	+2.1	+883	+1.2	+2,412	+3.4
Georgia	108,049	109,582	109,924	111,474	+1,533	+1.4	+1,875	+1.7	+3,425	+3.2
Florida	142,884	143,618	155,301	156,085	+734	+0.5	+12,417	+8.7	+13,201	+9.2
East South Central	359,764	366,838	365,844	372,971	+7,074	+2.0	+6,080	+1.7	+13,207	+3.7
Kentucky	85,722	85,870	86,047	86,189	+148	+0.2	+325	+0.4	+467	+0.5
Tennessee	98,049	98,853	100,401	101,216	+804	+0.8	+2,352	+2.4	+3,167	+3.2
Alabama	89,225	91,166	93,007	95,022	-1,941	+2.2	+3,782	+4.2	+5,797	+6.5
Mississippi	86,768	90,949	86,389	90,544	+4,181	+4.8	-379	-0.4	+3,776	+4.4
West South Central	483,304	487,933	485,305	489,841	+4,629	+1.0	+2,001	+0.4	+6,537	+1.4
Arkansas	52,670	53,830	51,721	52,855	+1,160	+2.2	-949	-1.8	+185	+0.4
Louisiana	122,477	125,507	121,114	124,100	+3,030	+2.5	-1,363	-1.1	+1,623	+1.3
Oklahoma	59,025	59,041	61,408	61,419	+16	-	+2,383	+4.0	+2,394	+4.1
Texas	249,132	249,555	251,062	251,467	+423	+0.2	+1,930	+0.8	+2,335	+0.9
West	954,071	948,497	956,783	951,139	-5,574	-0.6	+2,712	+0.3	-2,932	-0.3
Mountain	227,079	225,620	231,428	229,917	-1,459	-0.6	+4,349	+1.9	+2,838	+1.2
Montana	20,167	20,025	20,018	19,875	-142	-0.7	-149	-0.7	-292	-1.4
Idaho	21,118	20,992	21,474	21,344	-126	-0.6	+356	+1.7	+226	+1.1
Wyoming	9,542	9,472	9,526	9,455	-70	-0.7	-16	-0.2	-87	-0.9
Colorado	54,307	54,029	54,741	54,457	-278	-0.5	+434	+0.8	+150	+0.3
New Mexico	32,817	32,639	32,486	32,308	-178	-0.5	-331	-1.0	-509	-1.6
Arizona	47,707	47,344	49,768	49,385	-363	-0.8	+2,061	+4.3	+1,678	+3.5
Utah	29,907	29,663	31,966	31,703	-244	-0.8	+2,059	+6.9	+1,796	+6.0
Nevada	11,514	11,456	11,449	11,390	-58	-0.5	-65	-0.6	-124	-1.1
Pacific	726,992	722,877	725,355	721,222	-4,115	-0.6	-1,637	-0.2	-5,770	-0.8
Washington	80,785	79,987	81,205	80,396	-798	-1.0	+420	+0.5	-389	-0.5
Oregon	53,061	52,804	53,405	53,142	-257	-0.5	+344	+0.6	+81	+0.2
California	563,150	560,385	560,998	558,229	-2,765	-0.5	-2,152	-0.4	-4,921	-0.9
Alaska	6,227	6,178	6,172	6,123	-49	-0.8	-55	-0.9	-104	-1.7
Hawaii	23,769	23,523	23,575	23,332	-246	-1.0	-194	-0.8	-437	-1.8

- Zero or rounds to less than ± 0.05 percent.
[1]Distribution differs from official entitlements published by the Office of Revenue Sharing partly because the official entitlements exclude money held in reserve.
[2]Pertains to the general population factor and the urbanized population factor.

Table 6.7

(adopted from Siegel (1975b,p 28); numbers are in thousands)

GRS Allocations to States Based on 1970 Census Reported Data and 1970 Census Data Corrected by a Modified Synthetic Method

§6.3 Substate Errors in Allocation

Tables 6.8 and 6.9 present estimates of the moments of errors in allocations in EP's 1 and 6 for county areas in New Jersey. Restricting the scope of this part of the analysis to just one state was motivated by the large amounts of computing required by the analysis of each state. New Jersey was picked because it was one of the two states analyzed by Strauss and Harkins (1974). Their results will be discussed below. Note that the county areas under consideration do not refer to the total allocation within a county area, but to the amount designated for a county area at the final round of the "inner" iterations in the substate allocation algorithm (see §5.1).

County Area	Allocation	Bias POP(0)	Bias STANDARD	Bias POP(2)	Standard Deviation STANDARD
Atlantic	1,956 *	17.8	37.5	57.0	18.6
Bergen	4,469	40.7	171.5	301.9	143.
Burlington	1,716	15.6	49.0	82.2	70.9
Camden	3,370	30.7	82.7	134.3	152.
Cape May	665 *	6.1	17.8	29.5	6.32
Cumberland	1,078	9.8	21.7	33.5	62.3
Essex	10,420 *	94.8	95.5	95.1	98.9
Gloucester	1,187	10.8	35.3	59.6	53.3
Hudson	6,792 *	61.8	167.2	272.0	64.5
Hunterdon	436	4.0	17.6	31.2	22.1
Mercer	2,669	24.3	44.8	64.9	125.
Middlesex	4,091	37.2	145.9	254.1	160.
Monmouth	3,254	29.6	95.8	161.7	124.
Morris	2,039	18.5	80.3	141.8	72.7
Ocean	2,318	21.1	88.7	156.2	93.4
Passaic	4,019	36.6	100.9	164.9	173.
Salem	501	4.6	9.6	14.5	30.1
Somerset	994	9.0	36.8	64.4	41.1
Sussex	724	6.6	30.8	55.0	38.4
Union	3,432	31.2	86.1	140.6	123.
Warren	615	5.6	25.6	45.5	31.8
TOTAL	56,745	516.3	1,441.0	2,360.0	

Table 6.8

<u>Moments of Errors in New Jersey County Area Shares, EP 1</u>

(all figures in thousands of dollars)

* denotes PCLS-constrained allocation

In EP 1 (see Table 6.2) the upward bias in New Jersey's allocation incfeases as the population undercoverage rates increase (from Model POP(0) to POP(2)). This bias in the state's allocation is the dominant bias in the county areas' allocations, each of which is also biased upward in EP 1 (see Table 6.8). Note however that while the bias for each of the other county areas increases substantially with the undercoverage rates, the bias in Essex's allocation barely changes. To understand this, realize that since Essex County was PCLS-constrained in EP 1 the relative bias in its allocation is (from (5.35))

$$\mu_Y + \mu_P(j) \tag{6.1}$$

while the relative bias for an unconstrained county area is (from (5.32))

$$\mu_Y - 2\,\mu_C(j) \quad . \tag{6.2}$$

In fact, the part of the relative bias μ_Y for New Jersey that arises from undercoverage is roughly equal in magnitude to, but of opposite sign from, $\mu_P(j)$ for Essex County. Thus (6.1) varies only slightly as λ varies in POP(λ) and maintains the value 0.0092 for λ ranging from 0 to 2. On the other hand, the small sizes of $\mu_C(j)$ compared to μ_Y for unconstrained counties cause (6.2) to be positive for these counties as λ varies in POP(λ). In the other constrained counties $\mu_P(j)$ has the same sign as μ_P for New Jersey and so the bias in allocation increases as λ varies in POP(λ).

This phenomenon of a bias in the total allocation to the state causing the biases in county area allocations to all or nearly all having the same sign was also observed by the Stanford Research Institute (1974a, pp.D-20,21) for allocations in California. Their findings are reproduced in Table 6.10. This pattern would not be anticipated in states with small absolute bias in total allocation but large heterogeneity of substate undercoverage and money income relative bias rates.

Strauss and Harkins (1974) analyze errors in allocations within New Jersey and Virginia that arise from population undercoverage. Their approach resembles that of Siegel (1975b) in using synthetic estimates of undercount by age, race, and sex to calculate "adjustments" to the data and allocations. To estimate errors in per capita income estimates Strauss and Harkins assume that the uncounted people in a given race-sex category had incomes similar to their enumerated counterparts. Underreporting of income is not considered.

In estimating adjustments (or equivalently, errors) in allocations to counties in New Jersey and Virginia, Strauss and Harkins apparently did not consider errors in the state allocations. For New Jersey, which is our focus here, this is not too important since the relative bias in New Jersey's allocation is believed to be small under the models used by Strauss and Harkins. Their method of adjusting the data for undercount would, if applied at the state level, in effect be similar to the basic synthetic method applied to population only.* Under this model the adjustment for New Jersey would be small (see Table 6.6, col.7).

The results of Strauss and Harkins are not directly comparable with those of the present study because the underlying error models differ. We note their finding that in New Jersey in EP 1 (Strauss and Harkins 1974, Table 5) 17 counties had a positive relative bias (16 between .010 and .049 and 1 between .000 and .009) and 4 had a negative relative bias in allocation (between -.011 and -.050). This suggests that, at least in some states, there may be a heterogeneity in the signs of the biases in county area allocations. We may hypothesize however, noting the patterns observed earlier, that this latter pattern will be the exception and not the rule.

---

* This is because the differential effects of synthetic adjustment to state per capita income estimates would be small.

County Area	Allocation	Bias STANDARD	Standard Deviation STANDARD	POP-UP	PCI-UP
Atlantic	4,910 *	61.8	81.1	52.8	52.8
Bergen	10,250	308.5	354.	351.	285.
Burlington	5,090	117.2	211.	210.	184.
Camden	8,428	164.3	398.	397.	359.
Cape May	1,768 *	35.6	29.2	19.0	19.0
Cumberland	2,737	45.5	153.	152.	142.
Essex	22,169	-143.3	956.	951.	842.
Gloucester	3,144	75.3	138.	138.	122.
Hudson	15,581 *	260.8	257.	167.	167.
Hunterdon	1,262	40.9	58.6	58.3	52.6
Mercer	5,748	77.7	293.	292.	268.
Middlesex	10,008	281.8	422.	419.	369.
Monmouth	7,890	185.4	317.	315.	273.
Morris	5,586	174.6	221.	219.	189.
Ocean	6,591	209.9	270.	268.	234.
Passaic	8,683	167.5	396.	394.	354.
Salem	1,559	23.2	80.2	79.9	73.6
Somerset	2,545	74.2	110.	110.	97.0
Sussez	2,303	79.0	112.	111.	101.
Union	7,726	146.2	312.	310.	269.
Warren	1,754	58.0	86.2	85.9	78.4
Total	135,732	2,444.0			

Table 6.9

Moments of Errors in New Jersey County-Area Shares, EP 6

(all figures in thousands of dollars)

* denotes PCLS-constrained allocation

County	1970 Population		Under-enumeration Rate (percent)	Dollar Amount Allocated		Change in Allocation	
	Prior to Adjustment	After Adjustment		Prior to Adjustment	After Adjustment	Dollar Amount	Percent
Alameda	1,071,446	1,105,094	3.0%	$ 21,957,849	$ 21,974,447	$ 16,598	0.1%
Alpine	484	492	1.6	15,248	15,095	-153	-1.0
Amador	11,821	12,054	1.9	303,869	304,099	230	0.1
Butte	101,969	103,944	1.9	2,503,067	2,504,959	1,892	0.1
Calaveras	13,585	13,835	1.8	424,007	424,328	321	0.1
Colusa	12,430	12,739	2.4	349,891	350,155	264	0.1
Contra Costa	555,805	569,417	2.4	8,909,216	8,915,951	6,735	0.1
Del Norte	14,580	14,823	1.6	426,329	426,651	322	0.1
El Dorado	43,833	44,566	1.7	1,380,924	1,367,299	-13,625	-1.0
Fresno	413,329	427,495	3.3	12,739,314	12,748,944	9,630	0.1
Glenn	17,521	17,875	2.0	491,599	491,971	372	0.1
Humboldt	99,692	101,443	1.7	2,452,255	2,454,109	1,854	0.1
Imperial	74,492	77,847	4.3	2,346,813	2,388,370	41,557	1.8
Inyo	15,571	15,849	1.8	353,863	354,130	267	0.1
Kern	330,234	340,078	2.9	10,403,767	10,433,698	29,931	0.3
Kings	66,717	68,665	2.8	2,101,867	2,106,663	4,796	0.2
Lake	19,548	19,918	1.9	615,844	611,090	-4,754	-0.8
Lassen	16,796	17,128	1.9	358,966	359,237	271	0.1
Los Angeles	7,041,980	7,275,461	3.2	156,532,661	156,650,986	118,325	0.1
Madera	41,519	42,817	3.0	1,308,024	1,313,639	5,615	0.4
Marin	206,758	210,788	1.9	2,236,120	2,237,810	1,690	0.1
Mariposa	6,015	6,119	1.7	189,497	187,733	1,764	0.9
Mendocino	51,101	52,023	1.8	1,552,868	1,554,042	1,174	0.1
Merced	104,629	107,851	3.0	3,296,225	3,308,902	12,677	0.4
Modoc	7,469	7,605	1.8	235,305	233,324	-1,981	-0,8
Mono	4,016	4,080	1.6	126,500	125,176	-1,324	-1.0
Monterey	247,450	254,687	2.8	5,143,787	5,147,675	3,888	0.1
Napa	79,140	80,606	1.8	1,504,697	1,505,744	1,137	0.1
Nevada	26,346	26,781	1.6	677,545	678,057	512	0.1
Orange	1,421,233	1,452,423	2.1	22,407,553	22,424,491	16,938	0.1
Placer	77,632	79,171	1.9	1,729,081	1,730,388	1,307	0.1
Plumas	11,707	11,930	1.9	322,153	322,397	244	0.1
Riverside	459,074	472,065	2.8	11,879,059	11,888,039	8,980	0.1
Sacramento	634,373	639,777	2.4	13,838,271	13,848,732	10,461	0.1
San Benito	18,226	19,020	4.2	370,946	371,226	280	0.1
San Bernardino	682,233	700,997	2.7	16,857,782	16,870,525	12,743	0.1
San Diego	1,357,854	1,391,769	2.4	25,628,428	25,647,801	19,373	0.1
San Francisco	715,674	736,628	2.8	18,534,030	18,548,040	14,010	0.1
San Joaquin	291,073	299,978	3.0	7,919,637	7,925,624	5,987	0.1
San Luis Obispo	105,690	108,131	2.3	3,240,400	3,242,849	2,449	0.1
San Mateo	557,361	570,674	2.3	6,693,786	6,698,846	5,060	0.1
Santa Barbara	264,324	271,077	2.5	5,486,179	5,490,326	4,147	0.1
Santa Clara	1,065,313	1,092,106	2.5	18,086,200	18,099,871	13,672	0.1
Santa Cruz	123,790	126,463	2.1	3,072,959	3,075,282	2,323	0.1
Shasta	77,640	79,043	1.8	1,726,949	1,728,254	1,305	0.1
Sierra	2,365	2,408	1.8	69,420	69,472	52	0.1
Siskiyou	33,225	33,891	2.0	913,812	914,503	691	0.1
Solano	171,989	176,525	2.6	3,813,032	3,815,914	2,882	0.1
Sonoma	204,885	208,847	1.9	5,661,968	5,666,248	4,280	0.1
Stanislaus	194,506	198,884	2.2	5,104,732	5,108,591	3,859	0.1
Sutter	41,935	42,853	2.1	945,240	945,955	715	0.1
Tehama	29,517	30,068	1.8	834,155	834,786	630	0.1
Trinity	7,615	7,741	1.6	239,904	237,496	-2,408	-1.0
Tulare	188,322	194,131	3.0	5,932,939	5,955,999	23,060	0.4
Tuolumne	22,169	22,624	2.0	554,460	554,879	419	0.1
Ventura	378,497	388,978	2.7	8,504,537	8,510,966	6,429	0.1
Yolo	91,788	94,080	2.4	2,307,640	2,309,384	1,744	0.1
Yuba	44,736	45,809	2.3	1,341,885	1,342,899	1,014	0.1

Table 6.10

(adopted from SRI (1975a.p D-21))

Changes in GRS Allocations to County Areas in California After Adjusting for Underenumeration

In EP 6 (see Table 6.9) the situation differs from EP 1. Here Essex county's allocation is not PCLS-constrained and the large negative bias in $\mu_C(j)$ dominates the positive bias μ_Y of the total state allocation (see (6.2)) so that the bias in Essex County's allocation is negative. The remainder of the county area allocation biases are all positive, supporting the hypothesis that in most states the vast majority of the biases in allocations to county areas will have the same sign as the state's bias.

As expected, the reductions in the standard deviations of the county-area allocations under alternative error models differed for PCLS-constrained and unconstrained county areas (Table 6.9, POP-UP and PCI-UP columns). The standard deviations for Atlantic, Cape May, and Hudson counties all improved substantially when population updating error was eliminated (POP-UP). Improving the per capita income updates (PCI-UP) brought no further reductions in the standard deviations for these county areas. The situation was the reverse for the other (unconstrained) counties: improving the population updates alone barely affected their standard deviations but improving the per capita income updates had great impact. This is as expected because the shares of the state allocation going to constrained county areas is determined by population only, while the shares to unconstrained county areas are determined almost entirely on the basis of per capita income and adjusted taxes.

The reader is reminded that the estimates of population biases and especially per capita income biases for substate areas are unsophisticated (see §3.6) and while believed to be indicative on the whole, they are not reliable estimates of bias for individual areas. In particular, the model of §3.6.5 assigns the substate per capita income biases magnitudes roughly equal to, but signs opposite from, the population biases. For an individual substate unit, this model neglects response biases in money income and overstates the effect of population undercoverage. While the resulting estimates are inappropriate indicators of the biases for individual areas, they do yield estimates of the sum of the mean absolute misallocations over places in a county or counties in a state. The variation in estimated undercoverage rates is thus used as a proxy for the variation in the response bias rates. It is stressed that to the extent that estimates of biases in per capita income influence the estimate of bias in the allocation to a sub-state unit, this latter estimate may be more suggestive than accurate.

To limit the amount of computations, I decided to estimate the moments of errors in subcounty allocations only for places (and municipalities) in one county area, Essex. These estimates appear in Tables 6.11 (EP 1) and 6.12 (EP 6). Essex County was chosen becasue Strauss and Harkins (1974) presented estimates of the effects of undercounts on subcounty allocations only for Essex county area. Furthermore, Essex County contains Newark, which has challenged the Treasury Department over biases in GRS allocations arising from population undercoverage. The ruling (in favor of Treasury) can be found in City of Newark, N.J. et al. v. W. Michael Blumenthal, Secretary of the Treasury, et al. no. 74-548 (D.D.C. Jan. 17, 1978).

It was argued in chapter 3 that the covariance between the relative updating errors in the per capita income of places was between .0000 and .0006. Note that a large positive covariance will reduce the variance of the allocations, which contains a term like

$$\begin{aligned} \mathrm{Var}(x_k - \sum_{\ell} w_\ell x_\ell) &= (1 - 2\, w_k)\, \sigma^2 + \sum_{\ell} w_\ell^2\, \sigma^2 \\ &\quad -2(1 - w_k)^2\, c + \sum_{\substack{\ell, m \neq k \\ \ell \neq m}} \sum w_\ell w_m c \end{aligned}$$

where x_k here represents the relative error in per capita income with $\sigma^2 = \mathrm{Var}(x_\ell)$ and $c = \mathrm{Cov}(x_\ell\, , \, x_m)$ $\ell \neq m$ and where the w_k are non-negative weights summing to unity. The covariance will, unless otherwise noted, be assumed to be .0006. Such an assumption is termed conservative because it effects an understatement of the measured benefit from improving the quality of the sub-county per capita income updates. For illustrative purposes, some calculations will be reported in the case of zero covariance. This model is labled PCI-COV(0) and is the same as STANDARD, except that per capita income updating errors for places are assumed uncorrelated.

As with the county area allocations there is a marked tendency for the allocations to places in Essex County to exhibit upward biases (see Tables 6.11, 6.12). However the upward bias in New Jersey's allocation is only partly responsible for this sub-county pattern. For example, in EP 6 the bias in the allocation to Essex County area was negative (Table 6.9) but the biases in places' allocations (Table 6.12)were still greatly positive. This is understood by realizing that Newark received the lion's share of the allocations to all places in Essex County and that because of population and per capita income biases, the relative bias in the allocation to Newark was -0.020 while that for Essex County was only -0.006. Thus the magnitude of the bias in Newark's allocation was large enough to offset the effect on most places of the negative bias in Essex County area's EP 6 allocation.

Whether the kind of dominating role played by Newark is common or exceptional is a subject for future research. The Stanford Research Institute's (1974a, appendix C) analysis of biases in allocations within counties seems to indicate that the Essex County situation is exceptional (Newark received roughly three fourths of the EP 1 allocation to the whole county area). If, as is hypothesized, this is in fact the case, then as a general rule it may be anticipated that biases in county-area allocations will drive the biases in subcounty allocations to (nearly) all the same sign. This would imply that intra-county differentials in population undercoverage and income underreporting are of only minor importance.

To see this, note that the level of inequity (as measured by the sum of absolute errors in allocation; see (1.16)) is rather insensitive to the way in which the bias in the county-area allocation passes down into the subcounty allocations, provided that all of the latter are biased in the same direction. Exceptions to this pattern have a good chance of occuring when the county area contains a dominating town (such as Newark) or when the bias in the county area allocation is very close to zero.

The figures in Table 6.11 and 6.12 do not agree well with those of Strauss and Harkins (1974, Table 7). This lack of concordance arises from the differences between the error models of Strauss and Harkins and those of the present study. In particular, Strauss and Harkins neglect response bias (income underreporting) in the per capita income estimates. The reader is reminded that although the present estimates of per capita income bias are not reliable on a place by place basis, they are believed indicative on the whole, i.e. they are believed reliable for estimating the sum of mean absolute misallocations to places.

Tables 6.13 and 6.14 give estimates for various entitlement periods and error models of the sum of the mean absolute deviations in allocations to county areas in New Jersey and to places in Essex County.

It is remarkable that the sum of the mean absolute misallocations in EP 1 to places in Essex County (Table 6.13) is smaller for model POP(.5) and POP(1) (i.e. STANDARD) than for POP(0). Here <u>decreasing</u> the undercoverage differentials <u>increases</u> the errors in the allocations! This apparent paradox arises because the bias in Newark's allocation changes sign under POP(λ) when λ increases from 0 to 1, so that the bias in Newark's allocation is zero from some λ between 0 and 1.*

---

* Recall that λ regulates the severity of undercoverage differentials.

To understand this, we note two facts. First, New Jersey's undercoverage rate is somewhat lower than the national average and its allocation bias is positive and increases moderately as λ increases (see Table 6.2). Second, Newark's undercoverage rate is much higher than New Jersey's and Newark's EP 1 allocation is proportional to its estimated proportion of New Jersey's population. Thus the bias in Newark's share of New Jersey's allocation is negative and rapidly grows more severe as λ increases. But the upward bias in New Jersey's allocation more than offsets this negative bias in Newark's share (when λ is small enough), so that the bias in Newark's total allocation is positive. As we can see from Table 6.11, when λ equals 1 (STANDARD), the bias in New Jersey's allocation is no longer large enough to offset the bias in Newark's share, and Newark suffers a negative bias. How small need λ be for Newark's bias to be positive? Linear interpolation shows that λ need only be as small as 0.94. That is, if population undercoverage were improved at least 6 percent, Newark's EP 1 allocation would be biased upward. At the critical value of λ (approximately .94), Newark has no bias in its allocation and the sum of mean absolute misallocations to places in Essex County is minimized. But for λ smaller than .94, Newark's bias is positive, and the sum of mean absolute misallocations is increased.

Place	Allocation	Bias POP(0)	Bias STANDARD	Bias POP(2)	Standard Deviation STANDARD
Belleville	133.1	1.2	8.0	14.8	16.8
Bloomfield	149.9	1.4	9.2	17.0	18.8
Caldwell	17.7	.2	1.1	2.0	2.46
E. Orange	349.3	3.2	19.6	36.1	43.2
Essex Falls	3.9 *	.0	.1	.2	.04
Glen Ridge	15.6	.1	1.0	1.8	2.17
Irvington	223.2	2.0	12.9	23.8	27.8
Montclair	85.8	.8	1.8	2.9	10.9
Newark	4,267.6 *	38.8	-2.5	-44.3	40.5
N. Caldwell	10.4 *	.1	.3	.5	.99
Nutley	81.6	.7	5.0	9.3	10.4
Orange	169.1	1.5	1.3	1.1	21.4
Roseland	7.7	.1	.5	.9	1.18
So. Orange	27.8	.3	1.6	3.0	3.66
Verona	23.2 *	.2	.7	1.3	.22
W. Caldwell	24.5	.2	1.6	2.9	3.31
W. Orange	119.5	1.1	7.4	13.7	15.1
Fairfield	22.8	.2	1.5	2.7	3.26
Total	5,732.8	52.2	71.1	89.5	

Table 6.11

Moments of Errors in Allocations to Places in Essex County, EP 1

(all figures in thousands of dollars)

* denotes PCLS-constrained allocation

Place	Allocation	Bias STANDARD	Standard Deviation PCI-COV(0)	STANDARD	POP-UP	PCI-UP
Belleville	363.5	12.9	50.3	45.5	45.5	43.1
Bloomfield	424.0	15.5	58.6	53.0	53.0	50.3
Caldwell	43.4	1.6	6.34	5.78	5.78	5.51
E. Orange	849.2	26.9	117.	106.	106.	100.
Essex Falls	9.1 *	.2	.243	.243	.977	.977
Glen Ridge	43.6	1.6	6.36	5.80	5.80	5.54
Irvington	543.6	18.2	74.7	67.6	67.5	64.0
Montclair	230.0	.4	32.1	29.1	29.0	27.6
Newark	9,072.5	-187.7	92.5	91.4	91.3	83.9
N. Caldwell	24.9 *	.5	.411	.411	.267	.267
Nutley	239.2	8.7	33.3	30.1	30.0	28.5
Orange	430.9	-3.4	59.6	53.9	53.9	51.1
Roseland	42.1	1.6	6.51	6.00	6.00	5.76
So. Orange	72.8	2.5	10.4	9.39	9.39	8.93
Verona	57.7	2.1	8.20	7.44	7.43	7.07
W. Caldwell	57.8	2.2	8.30	7.54	7.54	7.18
W. Orange	273.6	10.1	38.1	34.4	34.4	32.6
Fairfield	72.8	2.8	10.8	9.87	9.87	9.43
Total	12,850.3	-83.2				

Table 6.12

Moments of Errors in Allocations to Places in Essex County, EP 6

(all figures in thousands of dollars)

* denotes PCLS-constrained allocation

Error Model		Sum of Mean Absolute Errors (millions of dollars)	Sum of Mean Absolute Errors as % of EP Allocation
EP 1	POP(0)	1.45	2.56
	POP(.5)	1.63	2.87
	STANDARD	1.89	3.33
	POP(1.5)	2.21	3.91
	POP(2)	2.59	4.56
	STINCTX	1.70	2.99
	ADJTAX(.5)	1.69	2.97
	ADJTAX(0)	1.48	2.60
EP 6	STANDARD	4.87	3.59
	POP-UP	4.80	3.54
	PCI-UP	4.44	3.27
	ADJTAX(.5)	4.25	3.13
	ADJTAX(0)	3.51	2.59

Table 6.13

Sums of Mean Absolute Errors in Allocation to County Areas in New Jersey

EP 1 Allocation = $56.75 million

EP 6 Allocation = $135.73 million

Error Model		Sum of Mean Absolute Errors (millions of dollars)	Sum of Mean Absolute Errors as % of Total EP Allocation
EP 1	POP(0)	.191	3.33
	POP(.5)	.184	3.22
	STANDARD	.190	3.32
	POP(1.5)	.209	3.64
	POP(2)	.238	4.15
	STINCTX	.192	3.35
	ADJTAX(.5)	.161	2.80
	ADJTAX(0)	.113	1.98
EP 6	STANDARD	1.134	8.82
	POP-UP	1.133	8.81
	PCI-UP	1.06	8.22
	ADJTAX(.5)	.89	6.91
	ADJTAX(0)	.54	4.19

Table 6.14

Sums of Mean Absolute Errors in Allocation to Places in Essex County

EP 1 Allocation = \$56.75 million

EP 6 Allocation = \$135.73 million

§6.4 Benefit Analysis

The benefit from improving the quality of the data elements will be measured with the loss function $L(\underset{\sim}{f},\underset{\sim}{\theta})$ discussed in §1.3 and §1.4. For simplicity (and with negligible loss in accuracy) errors in non-contiguous adjustments will be ignored, so that for EP's 1-7 GRS is treated as a fixed-pie allocation process. The loss function thus has the form

$$L(\underset{\sim}{f},\underset{\sim}{\theta}) = \frac{a-b}{2} \sum_{\ell} | f_{\ell} - \theta_{\ell} |$$

where $\underset{\sim}{f}$ and $\underset{\sim}{\theta}$ are vectors of actual and optimal GRS allocations respectively, ℓ indicates a recipient, and $(a-b)/2$ is a constant. The summation is over all recipients and over one or more entitlement periods.

It is useful to think of ℓ as denoting the triple ijk where i,j and k refer respectively to the governments of a state, county and subcounty unit and $j = 0$ or $k = 0$ denotes a "null unit". Thus, for example f_{i00} and f_{ij0} refer respectively to the allocation to state i and to county government j. Note that $f_{i..}$ is the total allocation to state i, which is approximately $3f_{i00}$.

Now consider evaluating

$$\text{(6.3)} \qquad E \sum_{\ell} |f_{\ell} - \theta_{\ell}| - E \sum_{\ell} |f'_{\ell} - \theta_{\ell}|$$

where f and f' are random GRS allocations arising from two different models, for example STANDARD and POP(λ). Since the allocation to each state government is roughly 1/3 the allocation to the entire state, (6.3) can be rewritten as

(6.4) $$\frac{1}{3}\left(E \sum_i |f_{i\cdot\cdot} - \theta_{i\cdot\cdot}| - E \sum_i |f'_{i\cdot\cdot} - \theta_{i\cdot\cdot}|\right)$$

(6.5) $$+ \left(E \sum_{\ell \neq i00} |f_\ell - \theta_\ell| - E \sum_{\ell \neq i00} |f'_\ell - \theta_\ell|\right) .$$

Note that (6.4) can be estimated from Table 6.4.

The present analysis does not allow for accurate estimation of (6.5) because only one state, and in it only one county, have been examined in detail. Furthermore, Essex County is believed to be atypical because of the dominating presence of Newark.

To derive an estimate of (6.3) we proceed under the hypothesis that the sign of the bias in the total state allocation determines the signs of the biases in allocations to substate units. From the assumption that biases in substate units do not change sign under model POP(λ) as λ varies (Newark is clearly an exception) when f refers to STANDARD and f' to POP(λ) it is reasonable to estimate (6.3) by three times the estimate of (6.4).

Notice that an important implication of the hypothesis that signs of the biases in the total state allocation generally determine the signs of the biases in substate allocations is that biases in substate estimates of population and per capita income do not appreciably affect the total loss from errors in allocations (measured with this loss function).

The following examples present benefit-cost analyses for improving various data components. The criterion will be whether the expected benefit from data improvement exceeds the expected cost. If analysis shows the criterion to be satisfied by a data program we will say the calculated benefit <u>justifies</u> the cost. In this case we may conclude that the data program should be (or should have been) undertaken if the funds to carry out the program are available.

We can now at last turn our attention to the important question of whether more money should have been spent to reduce undercoverage in the 1970 census.

Example 6.1 (undercoverage)

In planning for the decennial censuses, the Bureau of the Census must make decisions about whether or not various coverage improvement programs should be undertaken. The decisions are made by comparing the anticipated costs of the programs with the expected gains in coverage.* For any given program, accurate assessment of the relation between expected costs and expected coverage is quite difficult. Inferences are made on the basis of pretests or special censuses carried out under conditions which of necessity differ from the decennial census.

One coverage improvement program not used for the 1970 census was the Nonhousehold Sources Coverage Improvement Program (hereafter NSCIP). This involves using lists of persons obtained independently of the census to check on within household underenumeration. Possible sources of lists include driver's license lists and membership lists of ethnic organizations. If a person on a nonhousehold list is found not to have been enumerated at the address on the list, the household is contacted to determine if the person was there on the census date; if so, an attempt is made to obtain identifying information.

We consider the question of whether the NSCIP should have been used in 1970. Of course it is too late to re-do the 1970 census, but the analysis is informative and may be useful in planning for future censuses.

Using the NSCIP could increase the coverage of the 1980 census by .9 million people from "difficult to enumerate" areas (Census 1978). The estimated cost is $30.3 million (FY 1978 dollars).

---

* Other benefits, such as better public relations, may also be considered. For example, public criticism of the Census Bureau could be minimized or at least mitigated. Confidence in the government would also be increased, or in Congressman Lehman's words: "...confidence of the people in the census data is transferred to the confidence of the people in their political process" (Congress 1973, p.2812).

For the present purposes it is necessary to estimate the expected cost of the NSCIP if it were used in 1970. It is estimated that to repeat the 1970 census (which cost \$222 million in 1970) in 1980 would cost \$529 million (FY 1978 dollars). Accordingly, we estimate the expected cost of using the NSCIP in 1970 to be \$12.7 million.

Accurate estimation of the undercoverage rates that would have occurred under the NSCIP would require careful consideration of the deployment of the NSCIP efforts. For illustrative purposes a method of estimation will be sketched. As this method is laborious to execute, it will not be applied here.

The method, which presumes an "optimal" deployment of NSCIP efforts, is now described. Let P_k = population of area k, A_k = undercoverage rate of area k in the 1970 census, and A = national undercoverage rate in 1970. If the NSCIP had been used, the national undercoverage rate would have been .83A. Let $\gamma_k = A_k - .83A$. One possible kind of "optimal" deployment is to put NSCIP effots only in those areas k for which $\gamma_k > 0$, and to allocate efforts among these areas proportionally to $P_k \gamma_k$.

Note that putting NSCIP efforts into areas for which $\gamma_k \leq 0$ has the undesireable consequence of increasing the variability of the undercoverage rates. From these assumptions, estimates of the resulting distribution of undercoverage rates can be made. Unfortunately this requires a fair amount of computation. A method less demanding of effort will be used instead.

We will assume that the effect of the undercoverage improvements on GRS allocations can be modelled as POP (λ) for some λ; recall POP (λ) was described in §6.2 above. Motivation is the ease in calculation of the changes in the expected sum of absolute misallocations under POP (λ), discussed earlier. Since the NSCIP efforts are not spread uniformly but are concentrated on difficult to enumerate areas, the relevant value of λ is less than .83(= 1-.17). A more realistic value is $\lambda = .65(\approx 1-(2)(.17))$.

From Table 6.4 the estimated value of (6.3) for EP 1 with $\lambda = .5$ is \$15.2 million. By the convexity noted earlier, an estimated lower bound of (6.3) for EP 1 with $\lambda = .65$ is \$10.64 million.

Assuming that the ratio of (6.3) to $\Sigma\theta_\ell$ is constant for all EP's 1-11, the reduction in the sum of expected absolute errors over all EP's 1-11 is calculated from Table 6.1 to exceed

$$(6.6) \qquad \$224 \times 10^6 = (\$10.64 \times 10^6) \frac{(55.73 \times 10^9)}{(2.65 \times 10^9)} .$$

The magnitude of the benefit corresponding to (6.6) depends upon the value of (a-b)/2. In this and later examples we assume (a-b)/2 = .01 (it was argued in §1.4 that this is a reasonable value). Then the estimated expected benefit from improved GRS allocations is \$2.2 million. Since this is 18% of the cost of the coverage improvement program, consideration of the other benefits would be necessary to justify the cost. These benefits include not only improved representation in Congress and state legislatures, more accurate data for state and local planning, and better estimates of other characteristics, such as unemployment, which rely on population as a base, but improved allocations under a whole host of other programs relying on population data to allocate funds.

A recent study for the Subcommittee on Census and Population (Congress 1978) cited over 100 programs that allocate funds on the basis of population data. In fact, in Fiscal Year 1975, General Revenue Sharing accounted for less than 1/5 of the allocations made under the ten largest federal programs that allocate funds using population or per capita income (Office of Federal Statistical Policy and Standards 1978). Since improved allocations under GRS alone would account for 18% of the cost of improved coverage from NSCIP, consideration of benefits from improved funds allocation under other programs should account for most (if not all) of the rest of the cost. More analysis would be useful. I believe that the remaining cost, not justified on the basis of improved allocation of funds, is small enough to be unarguably justified by uses other than fund allocations.

Example 6.2 (state individual income tax collections)

Recall that for EP's 1 and 2 the data elements for 1972 state individual income tax collections were based on projections rather than estimates of amounts collected (see §3.9). Since the large biases in this data may well have arisen out of imperfect methodology rather than inaccurate data, it is reasonable to speculate that more money spent on obtaining estimates would have produced a great reduction in error.

From Table 6.4 it is seen that the estimated reduction in the sum of expected misallocations if there was no error in this data element would be $\$10.5 \times 10^6$. Since EP 1 and EP 2 state allocations were identical, the expected benefit in improved allocations from making flawless projections would have been

$$(\frac{a-b}{2})\ (\$21 \times 10^6) \quad .$$

Since we are using $(a-b)/2 = .01$, the expected benefit from improved estimates here would be \$210,000. Using the convexity assumption, we can estimate that the benefit from scaling the biases and standard deviations of the relative errors in state individual income tax collections by a factor of λ would be at least

$$(1-\lambda)\ (\$210{,}000) \qquad 0 \leq \lambda \leq 1 \quad .$$

To determine an "optimal" value of λ, say λ^*, it is necessary to have in hand a <u>cost function</u> $C(\lambda)$ mapping values of λ to expected costs in dollars. Then λ^* is a value of λ between 0 and 1 which maximizes

$$(6.7) \qquad 210{,}000(1-\lambda) \; - \; C(\lambda) \quad .$$

Without knowledge of the cost function C no definitive conclusions about λ^* can be made. If research were done to estimate C, I would be surprised if λ^* were greater than 1. That is, spending more money to improve projections of state individual income tax collections would have been justified on the basis of more accurate GRS allocations.

In fact, λ^* may be too large because other benefits from improved projections, such as better budgeting and superior estimation technology applicable to projecting other series, are not being considered. This claim is proved in the Technical Appendix (see proposition 6.1).

Example 6.3 (timeliness of population estimates)

Suppose that by spending an expected amount τ revised rather than provisional estimates of July 1, 1972 state populations could have been used in EP's 3 and 4. From Table 6.5 the reduction in the sum of mean absolute misallocations in these EP's is estimated to be $900,000. The benefit if $(a-b)/2 = .01$ is thus $9,000. It is unlikely that τ is nearly so small as $9,000. Here, the quantification of other benefits, such as improved planning or monitoring of trends, would have to greatly exceed $9,000 for improved timeliness to be justified.

Example 6.4 (local government adjusted taxes)

Consider the expected benefit from reducing by 50% the variances of Fiscal Year 1970-71 adjusted taxes estimates for all substate units in New Jersey (see §3.11). From Table 6.14 it is apparent that the sum of mean absolute misallocations to places in Essex County would be reduced by $\$29{,}000 = \$(.190-.161) \times 10^6$ in EP 1. To estimate the reduction for all substate allocations in New Jersey in EP 1 we assume that the relative reduction in the expected sum of absolute misallocations is constant for all substate entities in New Jersey. Note that the relative reduction for places in Essex County was $5.1 \times 10^{-3} = \$29 \times 10^{3}/(\$5.73 \times 10^{6})$. Since the total substate allocation in each EP was 2/3 of the total allocation to the state, and the total allocation to New Jersey for EP's 1-3 was $\$259.9 \times 10^{6}$, the estimated reduction in the sum of mean absolute misallocations is $\$884 \times 10^{3} = (5.1 \times 10^{-3})(2/3)(\$259.9 \times 10^{6})$. If $(a-b)/2 = .01$, the estimated benefit from improved allocations in EP's 1-3 is roughly $9,000.

Since the variance in the estimates of adjusted taxes arose primarily as response variance, improving the EP 1-3 data should have a carry-over effect for later estimates because the local officials who prepare the estimates would be better trained.

Furthermore, if the improvement could be obtained by better printed instructions to local officials, the adjusted taxes estimates could improve nationwide at a cost only slightly greater than the cost of improvement for New Jersey.

Little is known about the expected cost of reducing the response variance of adjusted taxes by means such as better instructions (as opposed to costly audits). The magnitude of the benefits from improved GRS allocations suggests that the possibility of improving the data be explored.

The examples above illustrate a variety of benefit-cost situations. In example 1, consideration of only the improved allocations in GRS that would result from better population coverage in the 1970 census is not enough to justify improving the coverage. However, consideration of other benefits indicates that certain coverage improvement programs should have been undertaken. In example 2, the benefits from improved GRS allocations are by themselves sufficient to justify improving the estimates of 1972 state individual income tax data, assuming the cost of this improvement would not have been extreme.

In examples 3 and 4 the benefits from inproved GRS allocations are rather moderate. In example 3, it is believed unlikely that other benefits could be quantified to justify improving the timeliness of the population estimates in EP's 3 and 4. Example 4 suggests that the costs of reducing the response variance in adjusted taxes be explored, since the benefit can potentially greatly outweigh the costs.

§6.5 Summary of Findings

Some of the findings of the preceding sensitivity and benefit analyses are summarized below.

1. Improvement of undercoverage rates (more specifically, reduction of undercoverage differentials) improve the overall accuracy of the GRS allocations, as measured by the sum of mean absolute misallocations. Such undercoverage improvement increased the mean absolute misallocations to an atypical minority of the states because biases in other data were no longer offset.

2. Reduction of variances in data improved the accuracy of the GRS allocations for each recipient government.

3. The greatest improvement in GRS allocations to states would come from joint improvment of the population and income data series, with improvement of per capita income estimates second in beneficial effects, improvement of population estimates third, and improvement of total money income fourth. Of course, the various data improvements would cost different amounts. The difference in rankings of per capita income and total money income arises because underreporting of income dwarfs population undercoverage as an important source of error in estimation of total money income. When total money income is divided by population to yield per capita income, the errors in the estimates of total money income and population do not cancel but often compound.

4. Because GRS has a hierarchical structure (allocations to states are divided among county areas; allocations to county areas are divided among subcounty governments) the error in a state's allocation often will dominate the errors in the substate allocations. For example, overallocation to a state may cause all the county-area allocations to be too large although some of the county-area shares are too small.

In entitlement period 1, GRS allocations to all county areas were biased upward because New Jersey's total allocation was positively biased (caused by New Jersey's population undercoverage rate falling below the national average.) But the interactions of errors in the numerous data series may be complex: the upward bias in New Jersey's total allocation would increase or decrease if population undercoverage were less or more severe, but the bias in allocation to Essex county would remain stable.

In contrast, Newark suffered a small negative bias in its entitlement period 1 allocation because its population undercoverage was (we estimate) more severe than that of New Jersey as a whole. Here the downward bias in Newark's share of New Jersey's allocation was too large for upward bias in the latter to offset. But if population undercoverage had been just slightly less severe, then the bias in Newark's share of the state total would have been smaller and the bias in New Jersey's allocation would have dominated, giving Newark an upward allocation bias.

5. More money should have been spent to reduce undercoverage in the 1970 census. Our analysis showed that if the Nonhousehold Sources Coverage Improvement Program had been used in the 1970 census, the benefits would have outweighed the costs. Benefits were measured largely by considering improvements in the allocations of federal funds. To compare these benefits with costs of better data we took the position that it was worth spending \$X for data to reduce the expected misallocations by \$100X.

6. Benefits from improving the 1972 projections of state individual income tax collections would have outweighed the costs.

7. The possibility of improving the adjusted taxes data needs to be explored, since the benefits in terms of improved allocation can potentially greatly outweigh the costs.

Chapter 7 Policy Perspectives and Recommendations

§7.0 Introduction and Summary

The preceding analyses as well as additional considerations discussed below have important policy implications regarding (I) construction of allocation formulas, (II) data quality, and (III) needed research. These implications are discussed and recommendations are made (§7.1-§7.9).

The recommendations may be summarized as follows:

I. Construction of Allocation Formulas

1. Since conflicting or reinforcing aims of the legislators can effect a collinearity among the variables in an allocation formula, the formulas should, where practicable, be constructed to exploit this effect so the amount of required data is reduced.

2. In constructing allocation formulas, new and severe data requirements should be noted and avoided where feasible.

II. Data Quality

3. Priority should be given to minimizing errors in data that cause large errors in allocations, provided that individual errors lie within broad tolerance limits. Specifically, the recommended strategy is to ensure that individual errors lie within the tolerance limits, and then to minimize the sum of absolute errors in allocation.

4. In tiered (including single-tiered) allocation programs data with uniform bias rates (in sign and size) should be used to determine allocations within any given tier. Attempts should be made to keep the sum of variances of the data as small as possible, with uniformity of variances not an objective.

5. The decision of whether to adjust the data underlying GRS allocations for suspected biases should be largely based on careful, rigorous analysis. Criteria to be satisfied by the adjustments should be stated clearly. In making the decision, uses of the data other than for allocation of funds should be considered.

6. Coverage improvements for the 1980 census should be increased from the level of 1970 efforts. In order of decreasing priority, the efforts should be aimed at reducing the undercoverage rates to uniform levels over states, then counties, then subcounty areas.

7. Explicit benefit-cost analysis should be used by statistical agencies in determining the levels of accuracy appropriate for allocating funds. In tiered allocation programs, if the analysis shows that one tier dominates in causing inequities, then data for that level should receive priority for improvement.

III. Needed Research

8. More research should be done on formulating error models for the data. The analysis of General Revenue Sharing suggests that substate data errors are of secondary importance because they are dwarfed by errors in state data. Research should extend to other tiered allocation programs to see if the finding is generally applicable.

9. More research should be conducted to see how well allocation formulas reflect Congressional intent, both at the time of enactment and over time. The implications for data quality may be important and should be explored.

10. Benefit analyses should be performed for uses of data other than allocating funds--in particular for state and local planning.

§7.1 Redundancy of Allocation Formulas

The great amount of data required to drive GRS allocation formulas places a possibly unnecessary burden on the statistical agencies. Nathan et al. (1975,p.92ff) and Dresch (1976, p.58) note that many of the variables in the state-level formulas cancel each other out because of high correlations.* Thus the correlation between EP 1 allocations to states and the population of states was greater than .9806. This suggests the possibility that a simpler allocation formula, i.e. one requiring less data input, could have been found acceptable by Congress.

For example, a straightforward allocation formula** using only three data elements reproduces almost exactly the states' actual EP 1 allocations as calculated on the basis of seven data elements. The largest difference between the EP 1 allocations for states under the two formulas was only $7.3 million, approximately 5 percent of the state's actual allocation. The correlation between the sets of EP 1 allocations prescribed by the two formulas was 0.9992. The data elements were population, per capita income, and net state and local tax collections (discussed above in §3.9).

---

* "Correlation" here refers to the cosine of the angle between two vectors. Note that this is an uncentered variant of the conventional correlation coefficient.

** The formula may be written

$$\frac{100Y_i}{Y.} = .0472 + 91.6\,\frac{T_i}{T.} - 56.5\,\frac{P_iC_i}{(PC).} + 62.5\,\frac{P_i/C_i}{(P/C).} \;,$$

where for state i we set Y_i = allocation, T_i = net state and local tax collections, P_i = population, and C_i = per capita income. The subscript "." denotes summation, so $Y.$ = total allocation to all states and $(PC).$ = sum over all states of the product of population and per capita income.

The performance over time of this regression was not examined; this should not be regarded as a major objection , since the performance of the actual allocation formula over time was not considered very carefully at the time it was adopted.

On the other hand, there may be practical reasons for including collinear variables in a formula. For example, suppose that a formula containing several data elements resulted in allocations virtually identical to those produced by a formula depending on population data alone. Although a pure population formula would demand fewer statistical resources, its enactment might be politically impossible. As such a formula does not respond to any special problem of federal concern many might argue that the funds should be alternatively targeted to achieve some identifiable federal goals. Thus the multitude of formula components (in GRS) may rationalize the program, giving it at least the appearance of responding to special considerations (urbanized population, tax effort, "need",...) even if these special considerations do cancel each other out. But where such rationalization is not absolutely necessary, the number of variables in the formula should be kept to a minimum.

Recommendation 1: Since conflicting or reinforcing aims of the legislators can effect a collinearity among the variables in an allocation formula, the formulas should, where practicable, be constructed to exploit this phenomenon so the amount of required data is reduced.

§7.2 Data Burdens Imposed by Legislation

The GRS legislation requires population, per capita income, adjusted tax, and intergovernmental transfer data for more than 39,000 local governmental units. Roughly 29,000 (22,000; 15,000) of these have fewer than 2,500 (1,000; 500) residents. Despite their number, the total allocation to these areas is quite small. For example, the total allocation to the 15,000 localities with fewer than 500 residents is less than 1.5% of the total GRS allocation.

Making updates of the population and per capita income estimates for these small areas is a significant burden for the Census Bureau. If the allocation program had been structured differently, it is possible that the intended allocation of funds could still have been satisfactorily approximated but with a smaller data burden on the Census Bureau. For example, consortia of contiguous small areas might have been set up as GRS allocation units, say with total population at least 10,000. Allocated amounts would be calculated for these units, which would then subdivide the money amongst themselves in some manner, for example, in proportion to 1970 population. Estimating the updated population or per capita income of these consortia could be done more easily and with greater accuracy than for the individual small areas.

Recommendation 2: In constructing allocation formulas, new and severe data requirements should be noted and avoided where feasible.

§7.3 Minimize Large Errors in Allocation

Statistical Policy Working Paper 1: Report on Statistics for Allocation of Funds (Office of Federal Statistical Policy and Standards 1978, p.27) recommends that

> ...since data errors are inevitable and since statistical resources are necessarily limited, priority be given to minimizing the very large errors which may occur in data used for the allocation of funds...To the extent that error measurements are available for small...areas one should check that relative errors are no greater than a prespecified maximum, but one should not be over-concerned with small errors since their effect on the total distribution is relatively minor.

The phrasing is not quite correct. Priority should go not to minimizing large errors in the data, but to minimizing errors in the data that cause large errors in allocation.

In chapter 1, especially §1.3-§1.8, a variety of loss functions were studied for use as criteria to determine optimal deployments of resources for collecting and analyzing data used for allocating funds. The analysis showed that the most satisfactory loss function measured loss proportionally to the sum of absolute errors in allocation. This loss function should not be applied uncompromisingly, however, since extremely large individual inequities may result, with possibly severe political consequences. Efforts should also be made to prevent individual errors from becoming unacceptably large. The recommended approach ensures that individual errors lie within broad tolerance limits, and then minimizes the sum of absolute errors in allocation. (This recommendation must be balanced against other needs for this general purpose data; see recommendation 5, below.)

Recommendation 3: Priority should be given to minimizing errors in data that cause large errors in allocation, provided that individual errors lie within broad tolerance limits. Specifically, the recommended strategy is to ensure that the indiviudal errors lie within the tolerance limits, and then to minimize the sum of absolute errors in allocation.

§7.4 Uniform Biases, Non-uniform Variances

In the present revenue sharing formulas it is not errors in the data *per se* that cause inequity but the variations of relative errors in data about national averages. For this reason it is important that the data have uniform bias rates across the states.

Recall that in the five-factor or "House" allocation formula (see (4.1) above) "income tax amount" (I_i ; see §3.11) depended on state individual income tax collections (K_i) for some states and on federal individual income tax liabilities (L_i) for other states. For most entitlement periods (EP's) the bias rates of the estimates of K_i and L_i were similar, but in EP's 1 and 2 the rates were quite different and contributed significantly to biases in the allocations. The lesson here is that lack of uniformity in the data elements used in the formulas--use of different data elements for different states--can have bad results if the bias rates in the data elements are different. For variances the situation is different; the goal should be reduction of all variances rather than maintenance of all relative variances at the same level. (See the Technical Appendix for more discussion.)

The phenomenon of biases cancelling should be attempted within each level or "tier" of a hierarchical or "tiered" allocation program.* Presumably, recognition of this fact partially motivated Recommendation 7 of the Office of Federal Statistical Policy and Standards (1978, p.27):

> ...[In] tiered allocation programs comparable data should be used for allocation to states, but policy flexibility may be allowed for sub-state allocations... .

---

* Tiered allocation programs are those which calculate allocations in such a way that allocations to large areas (e.g. states) are determined first, then the large-area allocations are divided among smaller areas (e.g. counties). Possibly the smaller-area allocations are further divided among still smaller areas (subcounty units, etc.).

The preceding analysis leads to a sharpening and extension of this recommendation.

Recommendation 4: In tiered (including single-tiered) allocation programs data with uniform bias rates (in sign and size) should be sued for allocations within any given tier. Attempts should be made to keep the sum of variances of the data as small as possible, with uniformity of variances not an objective.

§7.5 Adjusting for Undercoverage or Other Suspected Biases

Concern over the effects of population undercoverage bias on the allocation of GRS funds has prompted a number of authors to recommend "adjusting" the population data for the suspected undercounts (Strauss and Harkins 1974; Keyfitz et al. 1978). Jacob Siegel (1975b) makes the point that if you are going to adjust population estimates for undercounts, you should also adjust per capita income estimates to account for underreporting of income (see §3.5).

It should be noted that unless the estimates of bias are relatively accurate, the result of adjustment could well be to increase the inequity in the allocations. For example, if the estimates of undercoverage for states are very poor, then it is possible that the sum of expected errors in allocation would be increased if population data were adjusted for undercount. For although we would be eliminating the estimated bias in the data, we would be increasing the random error (variance) of the data.

It is worth noting that fully adjusting state population estimates for undercounts estimated by the demographic method is equivalent to ignoring the decennial census estimates of state population.* It is hard to see how this extreme action could be justified.

---

* This is because the demographic estimates are obtained "independently" of the census; Siegel et al. (1977).

Thus if adjustment for population undercount (or income underreporting) is to be performed, consideration must be given to combining the different estimates, i.e. census counts and demographic estimates.

Various persons have raised the point that adjusting data is not new--price indexes and unemployment statistics are "seasonally adjusted" and the 1970 census counts were subjected to imputations for persons not counted but believed to be living in housing classified as "vacant". However, the kinds of adjustment proposed for population undercoverage differ from these in important respects. Seasonal adjustments are done not to correct deficiencies in the data, but to adjust for changes in the parameter being estimated. Although the imputations for vacancies were made to correct deficiencies in the data, these imputations were based on evidence "internal" rather than "external" to the decennial census. That is, the imputations were part of the census procedure, made on the basis of a review of a national sample of 15,000 housing units classified as vacant by the original enumerators (Census 1976d, pp.14-15). The proposed adjustments for undercoverage would be made after the census on the basis of external evidence. While the estimates of population adjusted for imputations were acceptable for use to determine composition of the electoral college, it is not clear whether estimates adjusted for undercoverage could be so used.

Even if the overall accuracy or equity is improved (expected sum of misallocations is reduced) by adjusting for undercoverage, there are other concerns. For example, if adjusting the revenue sharing estimates for suspected biases is viewed as "tampering" with the data, a "min-max" criterion like the following may be desireable: adjusting shall not increase the severity of any negative bias in allocation.

Decreases (because of adjustment) in allocations are permitted only for areas that are overallocated. This is a strict criterion and difficult to verify. But it is unlikely that adjusting state population estimates by estimates of undercoverage from

Siegel et al. (1977) would satisfy the criterion; see National Research Council (1978, p. 12). Of course, adjustments below the state level would be even more error-prone. For further development and other approaches see Fellegi (1980) and Spencer (1980).

Many uses of public data, specifically population data, are not related to "dividing a pie", whether in allocating funds or apportioning representation. These uses are sensitive to estimates of the total population, rather than proportion of population, living in an area. For these uses the benefits from adjustments for total undercount would not be significantly reduced by problems arising from inaccurate adjustments for differential undercoverage. For example, a place's eligibility for self-government is determined in part on the basis of total population of the place. Adjusting for total undercoverage could make eligible for self-government more places that <u>should</u> be eligible. There is no offsetting negative loss here to a place for which only 50% of the undercoverage was adjusted, while, say, all other places had 70% of their undercounts adjusted. This contrasts to the effects on GRS allocations, where differential rather than total undercoverage rates are significant.

Other uses of this sort, for which total population rather than proportional population is significant, include numerous planning and policymaking activities at the national, regional, state and local levels in the public sector (and also in the private sector) and a variety of less publicized uses. For example, quotas of liquor licenses (in New York State) are proportional to population. In political primaries in some states (e.g. Arizona) the amount that can be spent by a candidate is proportional to the total population.

Keyfitz (1979) proposes that use of adjusted population data be limited to allocation of funds. Once established, the practice of adjusting data is sure to spread to other, perhaps most, uses of population data. For this reason, the decision of whether to use adjusted data for the allocation of funds should be made with other uses in mind.

Recommendation 5: The decision of whether to adjust the data underlying GRS allocations for suspected biases should be largely based on careful, rigorous analysis. Criteria to be satisfied by the adjustments should be stated clearly. In making the decision, uses of the data other than for allocation of funds should be considered.

§7.6 Improving the Coverage of the Decennial Census

Example 6.1 in chapter 6 compared the costs and benefits of improving the coverage of the 1970 census. The coverage improvement program considered was the Nonhousehold Sources Coverage Improvement Program (not used in 1970 but under consideration for 1980). The costs of implementing the program in 1970 were estimated by adjusting the predicted costs in 1980 for inflation and other cost increases over the decade. The coverage improvement for 1970 was estimated on the basis of evidence and estimates of the expected improvement for 1980. Accurate estimation of the gains in coverage from a particular program is generally difficult because (i) the evidence underlying the estimation is derived from special censuses and pretests, whose circumstances necessarily differ from those of the decennial census; and (ii) the gains from different coverage improvement programs may not be additive: e.g., two programs may both extend coverage to substantially the same subpopulations. Estimation of the costs of the programs is also difficult.

Given these shortcomings, the conclusion reached in Example 6.1 was that greater coverage efforts (particularly the Nonhousehold Sources Coverage Improvement Program) should have been expended in the

1970 census because the benefits would have outweighed the costs.*
Benefits were measured by estimating the improvements in the allocations of federal funds. To compare these benefits with costs of improving data we considered that it was worth spending $X for data to reduce the misallocations of funds by $100X. On this basis, improvements in General Revenue Sharing allocations alone justified one-fifth of the cost of the Nonhousehold Sources Coverage Improvement Program, and consideration of other benefits more than justified the rest of the cost. Assuming that the benefits from the program in the 1980 census data will be comparable or greater than those from the program in the 1970 census data, we conclude that significantly greater coverage improvement efforts should be made for 1980 than were made in 1970.

The estimates (presented in §6.2) of the means and standard deviations of errors in actual GRS allocations give reason for concern. Since analysis (e.g.figure 6.1 above) indicates that the inequity from errors in GRS allocations rises rapidly as differential undercoverage rates increase, it is concluded that allowing the undercoverage rates to worsen would have severe consequences in terms of inequities in allocation. The indications that increased coverage improvement efforts in 1980 will be necessary merely to prevent the 1970 undercoverage rates from slipping therefore makes increased coverage effort imperative.

The importance of differential undercoverage suggests that the coverage improvement efforts be aimed at equalizing undercoverage

---

* The conclusion was based on the implicit assumption that the funds necessary to carry out the coverage improvement program were available. When only limited funds are available, then those programs yielding the highest net benefit should be conducted. In this case further analysis would be necessary to compare the benefits and costs of better coverage with the benefits and costs of alternative programs.

rates across the country. The greatest benefits will come from uniform undercoverage rates for states, then counties, then subcounty entities.

Recommendation 6: Coverage improvement efforts for the 1980 census should be increased from the level of 1970 efforts. In order of decreasing priority, the efforts should be aimed at reducing the undercoverage rates to uniform levels over states, then counties, then to subcounty areas.

§7.7 State Errors Vs. Substate Errors

The analyses of §6.3 and §6.4 suggest that, under the present levels of error in the data, substate data errors are of secondary importance because they are dwarfed by errors in the state level data. This conclusion stems from the finding that biases in the total state allocation dominated the effects of biases in substate data, so that the latter tended not to effect the direction of the bias in allocation. Biases in substate data generally were found to cause some local governments to have larger overallocations or underallocations, while the sum of absolute errors within the state remained fairly constant. This implies that effort should first go in reducing errors in state-level rather than substate data.

If the biases in state-level data were reduced, the substate biases would no longer be dominated and their reduction would have a relatively greater effect on reducing inequity. Similarly, if the substate data biases had been larger, the state-level biases would not have dominated and, as above, their reduction would have had a relatively greater effect on reducing inequity.

The policy implication here is that if efforts are to be expended to improve data, the general strategy should be to reduce the biases in the state-level data, at least to the point where errors in substate data have a greater relative impact on the total inequity. Actually, this is a description of what a good strategy is expected to be in general. This general strategy is recommended when explicit benefit-cost analyses are not practical.

It must be stressed that these conclusions are tentative, because they are based on examination of only a small number of GRS recipients and because our hard knowledge of substate data biases is rather weak. More research is needed, both in formulating error models for the data elements and in estimating the effects of errors in data on the allocations to more recipients.

Research should also extend to other tiered (or hierarchical) allocation programs to see if the finding is _generally_ applicable, namely that errors in the data used to determine allocations to the primary recipient unit (e.g. states) should be reduced before attempting to reduce errors in the data used to subdivide these allocations among the secondary recipient units (e.g. county units).

Recommendation 7: Explicit benefit-cost analyses should be used by statistical agencies in determining the levels of accuracy appropriate for allocating funds. In tiered allocation programs, if the analysis shows that one tier dominates in causing inequities, then data for that level should receive priority for improvement.

Recommendation 8: More research should be done on formulating error models for data. The analysis of General Revenue Sharing suggests that substate data errors are of secondary importance because they are dwarfed by errors in state data. Research should extend to other tiered allocation programs to see if the finding is generally applicable.

§7.8 Allocation Formulas as Approximations

The variables used in allocation formulas are not "exact measures" but rather "proxies" for the variables the legislators have in mind; see Bixby (1977). This lack of exactness has an important implication for data quality: the less exact are the variables in the allocation formulas, the less data quality is needed. Actually, this is a conjecture which can be proved in some special cases. A proof is given in the Technical Appendix.

This hypothesis is not new. In his lively On the Accuracy of Economic Observations O. Morgenstern (1963, pp.117-118) considers the precision of economic observations with regard to the needs of economic theory and concludes "the maximum precision of measurement needed is dependent upon the power and fine structure of the theory using the measurement."

If the hypothesis is true in complicated situations such as GRS (and its veracity in simple cases suggests it is) and if the allocation formulas do not well represent Congressional intent then benefit-cost analyses based on the benefits from improved allocations may be misleading. In particular they may suggest overinvestment in data quality. Theorem 7.1 in the Technical Appendix demonstrates that as the difference between the actual legislated formala and the intended formula grows larger, the optimal level of data quality decreases.

More research needs to be done to assess how much the legislated and intended formulas differ. Research should also be done to estimate the magnitude of the effects of this difference on the optimal data quality.

Recommendation 9: More research should be done to see how well allocation formulas reflect Congressional intent, both at the time of enactment and over time. The implications for data quality may be important and should be explored.

§7.9 Other Uses of Data

The data elements used in General Revenue Sharing have many other uses. These uses are affected by errors in the data, and determining appropriate levels of data quality requires that these consequences be considered.

A number of analyses of the benefits of the data quality were mentioned in Chapter 0. An important class of uses of data that needs further sensitivity analysis is policy-making. Planning activities at the state and local levels constitutes a particularly useful area for this kind of research.

In choosing an area for future research on the benefits of data quality the following considerations are important.

(i) Errors in data should have significant consequences in its uses.

(ii) These consequences should be reasonably well understood, so measurement of the effects of errors in data is possible.

(iii) The sensitivity analysis should be conducive to generalization, either to future repetitions of the uses or to similar kinds of uses.

(iv) Satisfactory measurement of the benefits from reducing the effects of errors in data should be feasible.

I believe that state and local planning activities satisfy criteria (i)-(iv). Furthermore, sensitivity analysis of these activities would be especially timely. For example, much effort is currently going into improving the estimation of population and income for small areas. Planning activities constitute a major use of these estimates.

Recommendation 10: Benefit analyses should be done for uses of data other than allocating funds--in particular for state and local planning.

Recommendations 1-10 are summarized above in §7.0.

Appendix A Tables of Biases in Data

The following tables provide estimates of various biases discussed in chapter 3.

$\tilde{A}_i$	= estimated population undercoverage rate for state i (§3.2)
$\tilde{A}_{ui}$	= estimated urbanized population undercoverage rate for state i (§3.3)
$\tilde{E}(m_i(1)/M_i(1))$	= estimated fraction of money income in state i not reported in 1970 Census (§3.5)
$\tilde{A}_{ij}$	= estimated population undercoverage rate for county j in state i (§3.4)
$\tilde{A}_{ijk}$	= estimated population undercoverage rate for locality k in county j in state i (§3.4)

State	$\tilde{A}_i$ (×100)	$\tilde{A}_{Ui}$ (×100)	$\tilde{E}(M_i(1)/M_i(1))$ (×100)
Alabama	2.0	2.1	- 5.0
Alaska	5.6	0.0	- 6.0
Arizona	5.0	4.6	- 4.0
Arkansas	4.7	5.0	-11.0
California	4.5	4.6	- 8.0
Colorado	2.3	2.3	- 2.0
Connecticut	0.7	0.7	-10.0
Delaware	3.4	3.3	-12.0
D.C.	5.2	5.2	-17.0
Florida	5.6	5.6	- 1.0
Georgia	4.9	5.0	- 8.0
Hawaii	5.5	5.6	- 9.0
Idaho	3.3	3.4	- 4.0
Illinois	1.9	2.2	-11.0
Indiana	1.3	1.7	- 8.0
Iowa	0.9	0.9	-12.0
Kansas	0.7	0.9	-12.0
Kentucky	2.7	2.7	- 7.0
Louisiana	2.8	2.9	-11.0
Maine	2.7	2.6	-10.0
Maryland	1.9	2.2	- 6.0
Massachusetts	0.6	0.7	-10.0
Michigan	2.1	2.3	- 8.0
Minnesota	0.0	0.0	- 7.0
Mississippi	2.0	1.6	-11.0
Missouri	2.6	3.1	- 6.0

Estimates of State-level Relative Biases in Population, Urbanized Population, and Money Income

(all entries in hundredths)

Table A1

State	$\tilde{A}_i$ (×100)	$\tilde{A}_{Ui}$ (×100)	$\tilde{E}(m_i(1)/M_i(1))$ (×100)
Montana	0.9	0.8	- 7.0
Nebraska	0.8	1.2	-15.0
Nevada	6.3	6.5	-10.0
New Hampshire	2.0	2.0	- 4.0
New Jersey	1.0	1.2	- 8.0
New Mexico	8.3	8.1	- 9.0
New York	1.8	2.0	-12.0
North Carolina	3.1	3.3	- 9.0
North Dakota	0.8	0.7	- 9.0
Ohio	1.4	1.6	- 8.0
Oklahoma	1.5	1.6	- 3.0
Oregon	1.5	1.6	- 8.0
Pennsylvania	0.9	1.2	- 9.0
Rhode Island	0.9	0.9	- 8.0
South Carolina	5.5	5.2	- 8.0
South Dakota	2.3	1.6	-14.0
Tennessee	3.7	4.0	- 6.0
Texas	4.9	4.9	- 7.0
Utah	-0.1	-0.2	- 1.0
Vermont	1.9	0.0	- 8.0
Virginia	2.9	2.9	- 4.0
Washington	2.1	2.4	- 6.0
West Virginia	4.6	4.6	- 7.0
Wisconsin	-0.6	-0.5	- 5.0
Wyoming	0.3	0.0	- 5.0
NATIONAL	2.5	2.7	- 8.0

Estimates of State-level Relative Biases in Population, Urbanized Population, and Money Income

(all entries in hundredths)

Table A1 (cont.)

County	$\tilde{A}_{ij}$
Atlantic	1.6
Bergen	0.4
Burlington	1.0
Camden	1.1
Cape May	0.9
Cumberland	1.4
Essex	2.6
Gloucester	0.9
Hudson	1.1
Hunterdon	0.4
Mercer	1.5
Middlesex	0.6
Monmouth	0.9
Morris	0.4
Ocean	0.5
Passaic	1.1
Salem	1.4
Somerset	0.5
Sussex	0.3
Union	4.3
Warren	0.3

Estimates of Undercoverage Rates in New Jersey Counties

(all entries in hundredths)

Table A2

Place or Municipality	$\tilde{A}_{ijk}$
Belleville	0.4
Bloomfield	0.4
Caldwell	0.3
East Orange	0.6
Essex Falls	0.3
Glen Ridge	0.3
Irvington	0.5
Montclair	2.4
Newark	3.6
North Caldwell	0.8
Nutley	0.4
Orange	3.0
Roseland	0.2
South Orange	0.5
Verona	0.4
West Caldwell	0.3
West Orange	0.3
Fairfield	0.2

Estimates of Undercoverage Rates in Places and Municipalities in Essex County, New Jersey

(all entries in hundredths)

Table A3

Appendix B

Determination of General Revenue Sharing Allocations

General Revenue Sharing allocations are determined according to data-based formulas. Application of the formulas is complicated, performed by computer. The essentials of the procedure are described below; more complete discussions appear in chapter 4, §4.1 (state allocations) and chapter 5, §5.1 (substate allocations. Descriptions of the various kinds of data input to the formulas are given in chapter 3.

The calculation of General Revenue Sharing allocations proceeds through four major stages. First, allocations to the 51 state-areas (the fifty states and the District of Columbia) are determined. Next each state-area's amount is split into two shares, a state-government share and a statewide local-government share. Third, each statewide local-government share is partitioned among all county areas. The fourth stage involves calculating each local jurisdiction's allotment of the total available for the county area containing the jurisdiction. Local jurisdictions include: county governments, township governments, Indian Tribal Councils, Alaska Native Villages, and the governments of municipalities and other places. The presence of maximum and minimum constraints causes the second, third, and fourth stages to be performed several times.

Two formulas determine the allocations to state-areas: a "5-factor" and a "3-factor" formula. Because the "5-factor" and "3-factor" formulas originated with the House and Senate respectively, they are also referred to as the House and Senate formulas. The allocation to state-area i is proportional to the larger of the "House" amount H_i and the Senate amount S_i given by

$$(1) \qquad H_i = \frac{35}{159}\left(\frac{P_i}{P.} + \frac{P_i/C_i}{(P/C).} + \frac{U_i}{U.}\right) + \frac{27}{159}\left(\frac{I_i}{I.} + \frac{E_i T_i}{(ET).}\right)$$

and

$$(2) \qquad S_i = \frac{P_i E_i / C_i}{(PE/C).} \quad .$$

Here

P_i = population

U_i = urbanized population

C_i = per capita income

L_i = Federal individual income tax liabilities

K_i = State individual income tax collections

I_i = median$\{.01L_i , .15K_i , .06L_i\}$ = income tax amount

$E_i = T_i/R_i$ = tax effort

T_i = net state and local tax collections

and R_i = personal income

represent data elements provided by the Bureau of the Census and other agencies of the Department of Commerce. These data are described in chapter 3. The notation "." symbolizes summation over the subscript, for example $P. = \Sigma P_i$. The portion of the total pie allocated to state-area i can be written as $X_i/X.$, where X_i = maximum $\{S_i , H_i\}$. The size of the total pie was essentially fixed under the original program (P.L.92-512).

The allocation to each state-area is split 1:2 between the state government and all local governments in the state.*

---

* Actually, this ratio can vary slightly from state to state because of the effect of maximum and minimum provisions.

The allocation to all local governments is called the "local share".

The local share is then divided among all county-areas proportionally by the product of the county-area's tax effort and population divided by per capita income. The tax effort of a county-area is defined as the ratio of all "adjusted" (i.e. non-school) taxes collected by the county and sub-county governments to the product of the county's population and per capita income. Observe that the population factors cancel, so that the proportion of the local share going to county area j is

$$(3) \qquad \frac{D_{ij} + D_{ij.}}{(C_{ij})^2} \quad \frac{1}{G_i}$$

where the data elements are

D_{ij} = adjusted taxes of county government j

D_{ijk} = adjusted taxes of subcounty government k in county j

P_{ij} = population of county-area j

C_{ij} = per capita income of county-area j

and

$$G_i = \sum_j \left(\frac{D_{ij} + D_{ij.}}{(C_{ij})^2}\right) .$$

Note that population does not enter explicitly in formula (3). The county-area allocations are constrained by maximum and minimum provisions, however. No county-area allocation, on a per capita basis, is permitted to exceed 145% or fall below 20% of 2/3 of the state-area allocation, taken on a per capita basis. The allocation to a county area so constrained is thus proportional to the fraction of the state-area population residing in the county area, P_{ij}/P_i .

The partitioning of a county area allocation among local jurisdictions in the county takes place in several stages. Each Indian tribe or Alaskan village with members residing in the county is allocated a fraction of the county-area allocation equal to its proportion of the county-area population. Next, the remainder of the county-area allocation is partitioned amoung the county government, the ensemble all township governments (if any), and the ensemble of all place and municipality governments proportionally by the respective amounts of adjusted taxes of the three types of governments. The total for township governments is then allocated among the individual townships, so that the share for township k' is

$$\frac{D_{ijk'}}{(C_{ijk'})^2} \; \frac{1}{G_{ijk}}$$

where $C_{ijk'}$ is the per capita income of township k' and G_{ij} equals the sum over all townships k of $D_{ijk}/(C_{ijk})^2$. The total for all place and municipality governments is partitioned analogously.

The following maximum and minimum provisions apply to all local governments:

(i) No local government's allocation may exceed, on a yearly basis, 50 percent of the sum of its revenue from adjusted taxes and intergovernmental transfers.

(ii) No local government may receive, on a per capita basis, more than 145% or less than 20% of 2/3 of the state-area allocation, taken on a per capita basis.

(iii) Any local government allocation of less than $200 a year shall be forfeited by the locality and given to the county government.

Technical Appendix

Details are provided below for several examples and assertions mentioned earlier in §1.6, §6.2, §6.4, §7.4, and §7.8.

§1.6 Example 1.3

We consider a function that satisfies (1.37) but not (1.43). Suppose $\lambda_2 > 6$, $\lambda_1 < 1/6$. For $x > 0$ of the form 12^m let

$$H(x) = \frac{x}{2} \quad \text{for odd integers } m,$$

$$H(x) = 3x \quad \text{for even integers } m,$$

and let $H(x)$ be defined by linear interpolation otherwise. For example, if $12^m < x < 12^{m+1}$ where m is an odd integer, then

$$H(x) = \left(\frac{12^{m+1} - x}{12^{m+1} - 12^m}\right)\left(\frac{12^m}{2}\right) + \left(\frac{x - 12^m}{12^{m+1} - 12^m}\right)(3\cdot 12^{m+1}) .$$

Notice $x/2 \le H(x) \le 3x$ so $n/6 \le H(nx)/H(x) \le 6n$ and (1.37) holds. However, (1.43) fails because for $n = 12$

$$\frac{H(12x)}{H(x)} = \frac{3\cdot 12^{m+1}}{(.5)12^m} = 72 \qquad m \text{ odd},$$

$$\frac{H(12x)}{H(x)} = \frac{(.5)12^{m+1}}{3\cdot 12^m} = 2 \qquad m \text{ even},$$

and therefore $H(nx)/H(x)$ does not tend to a limit.

*****

§1.6 Corollary 1.8

Under the hypotheses of theorem 1.7 and regularity condition

(1.43) $$\psi(n) = \lim_{x \to \infty} H(nx)/H(x)$$ exists for all positive integers n,

the function H satisfies

(1.44) $$\lim_{x \to \infty} H(rx)/H(x) = r \qquad \text{for all } r \geq 0 .$$

Proof: First observe that (1.43) holds for all positive rationals, not just positive integers n. Letting m be a fixed positive integer, notice

(T.1)
$$\lim_{x \to \infty} \frac{H(xn/m)}{H(x)} = \lim_{mx \to \infty} \frac{H(mxn/m)}{H(mx)}$$
$$= \lim_{mx \to \infty} \frac{H(nx)}{H(x)} \frac{H(x)}{H(mx)}$$
$$= \psi(n)/\psi(m) .$$

Thus $\psi(n/m) = \psi(n)/\psi(m)$, so (1.43) holds for artitrary positive rationals, in fact

(T.2) $$\psi(rs) = \psi(r)\psi(s)$$

where $r,s > 0$ are rational. Since H is monotone, ψ can be defined everywhere by right continuity, so that (T.2) holds for all positive r and s. It is well-known (Feller 1971, p. 276 top) that all solutions to (T.2) that are bounded in finite intervals (essential for (T.1)) have the form $\psi(s) = s^\rho$ for some $\rho > 0$. Since satisfaction of (T.1) also implies $\rho = 1$, the conclusion is proved.

*****

§6.2 Bias in income estimates: underreporting vs. undercoverage

At the end of §6.2 it was stated that underreporting of income dominates population undercoverage as the major source of bias in money income estimates. To see this, note that the relative bias in per capita income, μ_C, is roughly $\mu_M - \mu_P$ (see §3.16 for notation). If μ_M arose mainly from undercoverage it would have the same sign as μ_P so that $|\mu_C|$ would be smaller than μ_M. In fact, this is not the case. Table A1 in Appendix A shows that for 22 states

$$\tilde{\mu}_M(i;1) = \tilde{E}(m_i(1)/M_i(1) - m(1)/M(1))$$

and

$$\tilde{\mu}_P(i;1) = \tilde{A} - \tilde{A}_i$$

have opposite signs (where $\tilde{E}m(1)/M(1) = -0.08$ and $\tilde{A} = 0.025$).

§6.2 Consequences of neglecting classes of benefits in benefit-cost analysis

In example 6.2 of §6.4 it was suggested that if some kinds of benefits from data improvement were neglected in benefit-cost analysis, the analysis would indicate less data improvement than if the benefits were not neglected. The reader is referred back to example 6.2 for definitions of C, λ, and λ^*. We now show, for a special case, that λ^* may be too large when some benefits are neglected. Let $\Psi\beta(\lambda)$ measure the neglected benefits and let A = \$210,000. Then the expected benefit is

$$\Psi\beta(\lambda) + A(1 - \lambda) - C(\lambda) . \tag{6.8}$$

If $\lambda(\Psi)$ maximizes (6.8) for fixed Ψ, then λ is a decreasing function of Ψ. This statement is demonstrated by the following proposition.

<u>Proposition 6.1</u>

Assume that for fixed Ψ (6.8) has a unique maximum at $\lambda(\Psi)$. Suppose further that $\Psi \geq 0$, $\lambda'(\Psi)$ exists, $\beta' < 0$, $\beta'' < 0$, and $C'' > 0$ (where $'$ denotes differentiation).
Then λ is a decreasing function of Ψ.

<u>Proof</u>: Since $\lambda(\Psi)$ uniquely maximizes (6.8), we have

$$\Psi\,\beta'(\lambda(\Psi)) - A - C'(\lambda(\Psi)) = 0 .$$

Differentiation with respect to Ψ yields

$$\beta'(\lambda(\Psi)) + \Psi\,\lambda'(\Psi)\,\beta''(\lambda(\Psi)) - \lambda'(\Psi)\,C''(\lambda(\Psi)) = 0 ,$$

or equivalently,

$$\lambda'(\Psi) = \frac{-\beta}{\Psi\,\beta'' - C''} < 0 .$$

The conclusion is immediate.

*****

§7.4 Uniform bias and nonuniform variance

In §7.4 it was claimed that uniform bias rates should be used for allocations within any given tier of a single or multiple-tiered allocation program, and that uniform variance rates were not important. The following analysis clarifies the meaning of the recommendation and demonstrates its validity.

Let $Y_i = f(Z_i)/(\Sigma_1^n f(Z_j))$ and $\theta_i = f(A_i)/(\Sigma_1^n f(A_j))$ denote the actual and optimal fractions of the GRS pie allocated to recipient i and suppose that $Z_i = A_i + \varepsilon_i$ and also that the delta method (theorem 2.1) can be applied to Y_i. The expected loss from errors in allocation is taken to be proportional to $\Sigma\, E|Y_i - \theta_i|$. Roughly, to minimize expected loss, one wants to make $E|Y_i - \theta_i|$ and $\mathrm{Var}(Y_i)$ small for all i, subject to budget constraints. The following decomposition suggests an economical way to proceed. Application of the delta method shows that $Y_i - \theta_i$ is approximately

$$\theta_i\left(\varepsilon_i \frac{f'(A_i)}{f(A_i)} - \frac{\Sigma\, \varepsilon_j\, f'(A_j)}{\Sigma\, f(A_j)}\right)$$

with mean approximately

$$\theta_i\left(E[\varepsilon_i \frac{f'(A_i)}{f(A_i)}] - \frac{\Sigma E[\varepsilon_j\, f'(A_j)]}{\Sigma\, f(A_j)}\right) \tag{7.1}$$

and variance approximately

$$(7.2)\qquad \theta_i^2\left\{ \mathrm{Var}(\varepsilon_i)\left(\frac{f'(A_i)}{f(A_i)}\right)^2 + \sum_j \mathrm{Var}(\varepsilon_j)\left(\frac{f'(A_j)}{\sum_k f(A_k)}\right)^2\right\}$$

$$+\ \theta_i^2\left\{\sum_{k\neq j} \mathrm{Cov}(\varepsilon_j\ ,\ \varepsilon_k)\ \frac{f'(A_j)\ f'(A_k)}{[\sum_\ell f(A_\ell)]^2}\right.$$

$$\left. -\ 2\sum_j \mathrm{Cov}(\varepsilon_i\ ,\ \varepsilon_j)\ \frac{f'(A_i)}{f(A_i)}\ \frac{f'(A_j)}{\sum f(A_k)}\right\} .$$

Notice that (7.1) equals zero when $E\varepsilon_i = c\cdot f(A_i)/f'(A_i)$ no matter what value c assumes. For the (common) case where f' is nearly constant over the range of the A_j, this says that uniform relative biases in the data will cancel. On the other hand, no such cancelling can occur in (7.2) unless the ε_j have a degenerate probability distribution. In practice, the first line of (7.2) dominates, and it is only by reducing $\mathrm{Var}(\varepsilon_i)$ that $\mathrm{Var}(Y_i)$ can be reduced.

§7.8 Theorem 7.1

In §7.8 the following hypothesis was considered: the less exact are the variables in the allocation formulas, the less data quality is needed. We now prove this hypothesis in a special case. Let $\underset{\sim}{\theta}$ be the vector of allocations intended by Congress, and let $\underset{\sim}{X}_N$ be the vector of actual data-based allocations. The subscript N is a parameter of data quality (see §2.3, above). We assume that

(7.8) $\underset{\sim}{X}_N$ is multivariate normal with mean $\underset{\sim}{\mu}$ and covariance matrix ΣN^{-1}, where $\underset{\sim}{\mu} - \underset{\sim}{\theta} = \underset{\sim}{B}$.

Because the allocation formula does not exactly

represent Congressional intent, even under perfect data the intended allocations and actual allocations will differ ($\underset{\sim}{B} \neq \underset{\sim}{0}$).

Assume that the formula is fixed-pie, so that the sum of the allocations does not depend on the data, and thus $\mathrm{Var}(\sum_j X_{N,j}) = 0$. Using (1.26) let the loss from misallocations be measured by

$$S\,\Sigma|X_{N,j} - \theta_j| \;+\; T\,(\Sigma\, B_j)^+ \;+\; U\,(\Sigma\, B_j)^-$$

for some $S > 0$ and let the cost of obtaining data satisfying (7.8) be VN^γ for some positive V and γ . Then the risk $R(N,\underset{\sim}{B})$ can be expressed as

$$R(N,\underset{\sim}{B}) \;=\; E(S\,\Sigma|X_{N,j} - \theta_j| \;+\; T\,(\Sigma\, B_j)^+ \;+\; U\,(\Sigma\, B_j)^- \;+\; VN^\gamma. \tag{7.9}$$

The optimal level of data quality, N^* , is taken to be infimum of all non-negative N that minimize (7.9). Regarding N^* as a function of $\underset{\sim}{B}$ we have the following theorem which applies when some data must be collected.

<u>Theorem 7.1</u>

Assume (7.8) and (7.9) hold and let β be an open set such that $N^* > 0$ as $\underset{\sim}{B}$ varies over β . Then

$$\frac{\partial N^*}{\partial B_i} \begin{Bmatrix} > \\ < \end{Bmatrix} 0 \quad \text{if} \quad B_i \begin{Bmatrix} < \\ > \end{Bmatrix} 0 \tag{7.10}$$

for all $\underset{\sim}{B}$ in β . That is, for $\underset{\sim}{B}$ in β fixed except for the ith component B_i ,

N^* is a decreasing function of $B_i > 0$

and

N^* is an increasing function of $B_i < 0$.

<u>Proof</u>: By (2.25) we have

$$R(N,\underset{\sim}{B}) = S\cdot\Sigma\Big\{2\,\sigma_j\,N^{-1/2}\,\phi(N^{1/2}B_j/\sigma_j) + B_j(1 - 2\,\Phi(-N^{1/2}B_j/\sigma_j)) + T\,(\Sigma\,B_j)^{+} + U\,(\Sigma\,B_j)^{-} + V\,N^{\gamma}\,,$$

where σ_j is the jth diagonal element of Σ. By hypothesis, N^* is strictly positive and therefore satisfies

$$0 = \left.\frac{\partial R(N,\underset{\sim}{B})}{\partial N}\right|_{N^* = N}\,.$$

Differentiating, we obtain

$$\frac{\partial R}{\partial N} = S\cdot\Sigma\Big\{-\sigma_j\,N^{-3/2}\,\phi(N^{1/2}B_j/\sigma_j) - N^{-1/2}\,\frac{B_j^2}{\sigma_j}\,\phi(N^{1/2}B_j/\sigma_j) - 2\,B_j\,\phi(N^{1/2}B_j/\sigma_j)\,(\tfrac{-1}{2}\,N^{-1/2}B_j/\sigma_j)\Big\} + V\gamma N^{\gamma-1}$$

$$= -\,S\,N^{-3/2}\,\Sigma[\sigma_j\,\phi(N^{1/2}B_j/\sigma_j)] + V\gamma N^{\gamma-1}\,.$$

Let

$$F(N,\underset{\sim}{B}) = N^{-\gamma-1/2}\,\Sigma[\sigma_j\,\phi(N^{1/2}B_j/\sigma_j)]$$

and notice that N^* lies on the surface $F(N,\underset{\sim}{B}) = V\gamma/S$. Application of the chain rule (Apostol 1969) gives

$$\frac{\partial F}{\partial B_i} + \frac{\partial F}{\partial N^*}\,\frac{\partial N^*}{\partial B_i} = 0$$

and so

$$\frac{\partial N^*}{\partial B_i} = -\left(\frac{\partial F}{\partial B_i}\right) / \left(\frac{\partial F}{\partial N^*}\right) .$$

Taking derivatives, we have

$$\frac{\partial F}{\partial B_i} = -N^{-\gamma+1/2} \phi(N^{1/2}B_i/\sigma_i) B_i/\sigma_i \left\{ \begin{matrix} < \\ > \end{matrix} \right\} 0 \quad \text{if} \quad B_i \left\{ \begin{matrix} > \\ < \end{matrix} \right\} 0$$

and

$$\frac{\partial F}{\partial N^*} = - N^{-\gamma-3/2} \Sigma[\phi(N^{1/2}B_j/\sigma_j) \sigma_j^{-1} (\frac{N}{2} B_j^2 + (\gamma + \frac{1}{2}) \sigma_j^2)] < 0$$

and thus $\frac{\partial N^*}{\partial B_i} \left\{ \begin{matrix} < \\ > \end{matrix} \right\} 0$ according to $B_i \left\{ \begin{matrix} > \\ < \end{matrix} \right\} 0$.

*****

The theorem says that as the lack of exactness, $\underset{\sim}{B}$, in the way the allocation formula represents Congressional intent gets worse, the optimal level of data quality decreases. More research needs to be done to assess what realistic values of $\underset{\sim}{B}$ might be and to estimate the effect on the level of optimal data quality, N^* . Alternative probability models for $\underset{\sim}{X}_N$ and alternative Fisher-consistent loss structures need to be considered as well.

BIBLIOGRAPHY

Apostol, T. (1957) Mathematical Analysis. Reading, Mass.: Addison-Wesley Publishing Co., Inc.

Apostol, T. (1969) Calculus, Vol. II. 2nd Edition. Waltham, Mass.: Blaisdell Publishing Co.

Bishop. Y. M. M.; Fienberg, S. E.; and Holland, P. W. (1975) Discrete Multivariate Analysis. Cambridge, Mass.: MIT Press.

Bixby, Lenore E. (1977) Statistical Data Requirements in Legislation. Committee on National Statistics, National Research Council. Washington, D.C.: National Academy of Sciences.

Black, H. C. (1968) Black's Law Dictionary. 4th Edition. St. Paul, Minn.: West Publishing Co.

Bradford. D. F. et al. (1974) The Value of Improved Information Based on Domestic Distribution Effects of United States Agricultural Crops. NASA Report 74-2001-5(N75-12423#). Princeton, N. J.: Economics Incorporated.

Census, U. S. Bureau of the (1972a) Census of Population and Housing: 1970 Evaluation and Research Program PHC(E)-1, The Quality of Residential Geographic Coding. Washington, D. C.: U. S. Govt. Printing Office.

Census, U. S. Bureau of the (1972b) Governmental Finances in 1970-71. Series GF75. Washington, D. C.: U. S. Govt. Printing Office.

Census, U. S. Bureau of the (1972c) Quarterly Summary of State and Local Tax Revenue. Washington, D. C.: U. S. Govt. Printing Office.

Census, U. S. Bureau of the (1973a) Census of Population: 1970, Vol. 1, Characteristics of the Population. Parts 2-52. Washington, D. C.: U. S. Govt. Printing Office.

Census, U. S. Bureau of the (1973b) Current Population Reports. Series P-25. Washington, D. C.: U. S. Govt. Printing Office.

Census, U. S. Bureau of the (1973c) Current Population Reports. Series P-26. Washington, D. C.: U. S. Govt. Printing Office.

Census, U. S. Bureau of the (1973d) Governmental Finances in 1971-72. Series GF75. Washington, D. C.: U. S. Govt. Printing Office.

Census, U. S. Bureau of the (1974a) Census of Population and Housing: 1970 Evaluation and Research Program PHC(E)-4, Estimates of Coverage of Population by Sex, Race, and Age: Demographic Analysis. Washington, D. C.: U. S. Govt. Printing Office.

Census, U. S. Bureau of the (1974b) Current Population Reports. Series P-25. Washington, D. C.: U. S. Govt. Printing Office.

Census, U. S. Bureau of the (1974c) Current Population Reports. Series P-26. Washinton, D. C.: U. S. Govt. Printing Office.

Census, U. S. Bureau of the (1974d) Governmental Finances in 1972-73. Series GF75. Washington, D. C.: U. S. Govt. Printing Office.

Census, U. S. Bureau of the (1975a) Boundary and Annexation Survey 1970-1973. Report GE30-1. Washington, D. C.: U. S. Govt. Printing Office.

Census, U. S. Bureau of the (1975b) Current Population Reports. Series P-25. Washington, D. C.: U. S. Govt. Printing Office.

Census, U. S. Bureau of the (1975c) Governmental Finances 1973-74. Series GF75. Washington, D. C.: U. S. Govt. Printing Office.

Census, U. S. Bureau of the (1976a) Current Population Reports. Series P-25. Washington, D. C.: U. S. Govt. Printing Office.

Census, U. S. Bureau of the (1976b) Governmental Finances in 1974-75. Series GF75. Washington, D. C.: U. S. Govt. Printing Office.

Census, U. S. Bureau of the (1976c) U. S. Census of Population and Housing: 1970 Procedural History PHC(R)-1. Washington, D. C.: U. S. Govt. Printing Office.

Census, U. S. Bureau of the (1978) Memorandum from Charles Jones, Chief of Statistical Methods Division to Daniel B. Levine, Associate Director of Demographic Fields, et al., Subject: Unit Costs, Coverage Improvement Program. April 28, 1978.

Cochran, W. G. (1977) Sampling Techniques. 3rd Edition. New York: John Wiley & Sons, Inc.

Coleman, E. J. (1974) Development of the components of personal and money income for the states and counties. Pp.8-13 in U. S. Bureau of the Census, Census Tract Papers, Series GE-40, No. 10, Statistical Methodology of Revenue Sharing and Related Estimate Studies. Washington, D. C.: U. S. Govt. Printing Office.

Coleman, E. J. (1977) Personal income: some observations on its construction, uses and adequacy as a subnational income measure. Pp.29-33 in U. S. Bureau of the Census, Small-area Statistics Papers, Series GE-41, No. 4, Interrelationship Among Estimates ,Surveys, and Forecasts Produced by Federal Agencies. Washington, D. C.: U. S. Govt. Printing Office.

Congress, U. S. (1973) Bills to Extend and Amend the Elementary and Secondary Education Act of 1965, and for Other Purposes. Hearings before the House Committee on Education and Labor, 93rd Congress, 1st Session, 3 May 1973. Washington, D. C.: U. S. Govt. Printing Office.

Congress, U. S. (1977) Pretest Census in Oakland, California and Camden, New Jersey, Hearings before the Subcommittee on Census and Population of the Committee on Post Office and Civil Service, 95th Congress, 1st Session, 25 March and 16 May 1977. Washington, D. C.: U. S. Govt. Printing Office.

Congress, U. S. (1978) Use of Population Data in Federal Assistance Programs. Subcommittee on Census and Population, Committee on Post Office and Civil Service, House of Representatives, 95th Congress, 2nd Session. Washington, D. C.: U. S. Govt. Printing Office.

Cramer, Harold (1945) Mathematical Methods of Statistics. Princeton, N. J.: Princeton University Press.

Debreu, G. (1965) Theory of Value: An Axiomatic Analysis of Economic Equilibrium. Cowles Foundation Monograph, no. 17. New Haven, Ct.: Yale University Press.

Decanio, S. (1978) Economic Losses from Forecasting Error in Agriculture. Department of Economics, Yale University, New Haven, Ct.

Dommel, P. R. (1974) The Politics of Revenue Sharing. Bloomington, Indiana: Indiana University Press.

Dresch, S. (1976) Monthly Labor Review 99(12): 58-9.

Dugundji, J. (1966) Topology. Boston, Mass.: Allyn and Bacon.

Efron, B. and Morris, C. (1971) Limiting the risk of Bayes and empirical Bayes estimators--part 1: the Bayes case. Journal of the American Statistical Association 66, 807-15.

Fay, R. E. III and Herriot , R. (1979) Estimates of income for small places: an application of James-Stein procedures to census data. Journal of the American Statistical Association 74(366):269-277.

Fellegi, I. P. (1980) Should the census count be adjusted for allocation purposes?--equity considerations. To appear in U. S. Bureau of the Census, Proceedings of the Conference on Census Undercount, Arlington, Va. 1980.

Feller, W. (1968) An Introduction to Probability Theory and Its Application, Volume I. 3rd Edition, New York: John Wiley & Sons.

Feller, W. (1971) An Introduction to Probability Theory and Its Application, Volume II. New York: John Wiley & Sons.

Ferreira, J. (1978) Identifying equitable insurance premiums for risk classes: an alternative to the classical approach. Pp.74-120 in Division of Insurance, Commonwealth of Massachusetts, Automobile Insurance Risk Classification: Equity and Accuracy: Boston: Massachusetts Division of Insurance.

Firth, R. (1952) Ethical absolutism and the ideal observer. Philosophy and Phenomenological Research 12: 336-341.

Fishburn, P. C. (1968) Utility theory. Management Science 14: 335-378.

Fisher, R. A. (1959) Statistical Methods and Scientific Inference, 2nd Edition. Edinburgh: Oliver and Boyd.

Friedman, M. and Savage, L. J. (1948) The utility analysis of choices involving risk. The Journal of Political Economy 56: 279-304.

General Accounting Office (1975) Adjusted Taxes: An Incomplete and Inaccurate Measure for Revenue Sharing Allocations. Washington, D. C.: U. S. Govt. Printing Office.

Gonzalez, M. E. (1974) Use and evaluation of synthetic estimates. Pp.46-50 in U. S. Bureau of the Census, Census Tract Papers, Series GE-40, No. 10, Statistical Methodology of Revenue Sharing and Related Estimate Studies. Washington, D. C.: U. S. Govt. Printing Office.

Hansen, M. H.; Hurwitz, W. N.; and Bershad, M. (1961) Measurement errors in censuses and surveys. Bulletin of the International Statistical Institute 38,2: 359-374.

Harsanyi, J. C. (1955) Cardinal welfare, individualistic ethics, and interpersonal comparisons of utility. Journal of Political Economy 61: 309-321.

Hayami, Y. and Peterson, W. (1972) Social returns to public information services: statistical reporting of U. S. farm commodities. The American Economic Review 62: 119-130.

Herriot, R. A. (1974) Preparations of final revenue sharing estimates of money income for political jurisdictions. Pp.18-25 in U. S. Bureau of the Census, Census Tract Papers, Series GE-40, No. 10, Statistical Methodology of Revenue Sharing and Related Estimate Studies. Washington, D. C.: U. S. Govt. Printing Office.

Herriot, R. A. (1978) Updating per capita income for General Revenue Sharing. Pp.8-15 in U. S. Bureau of the Census, Small-Area Statistics Papers, Series GE-41, No. 4, Interrelationships Among Estimates, Surveys, and Forecasts Produced by Federal Agencies. Washington, D. C.: U. S. Govt. Printing Office.

Hill, R. B. and Steffes, R. B. (1973) Estimating the 1970 Census Undercount for State and Local Areas. National Urban League Data Service, Washington, D. C.

Hotelling, H. (1938) The general welfare in relation to the problems of taxation and of railway and utility rates. Econometrica 6: 242-269.

Hsu, P. L. (1949) The limiting distribution of functions of sample means and application to testing hypotheses. Pp.359-402 in J. Neyman, ed., Proceedings of the Berkeley Symposium on Mathematical Statistics and Probability. Berkeley, Ca.: University of California Press.

Internal Revenue Service (1973) Statistics of Income--1971, Individual Income Tax Returns. Washington, D. C.: U. S. Govt. Printing Office.

Jabine, T. B. and Schwartz, R. E. (1974) Use of loss functions to determine sample size in the social security administration. Proceedings of the Social Sciences Section of the American Statistical Association, 103-110.

Jabine, T. B. (1977) Equity in the allocation of funds based on sample data. Pp.2-8 in U. S. Bureau of the Census, Small-Area Statistics Papers, Series GE-41, No. 3., Conference on Small-Area Statistics, Boston, 1976. Washington, D. C.: U. S. Govt. Printing Office.

Joint Committee on Internal Revenue Taxation, Staff of the (1973) General Explanation of the State and Local Fiscal Assistance Act and the Federal-State Tax Collection Act of 1972. Washington, D. C.: U. S. Government Printing Office.

Keeney, R. L. and Raiffa, H. (1976) Decisions with Multiple Objectives. New York: John Wiley & Sons.

Keyfitz, N. (1979) Information and allocation: two uses of the 1980 census. American Statistician 33: 45-50.

Kramer, G. H. and Klevorick, A. K (1973) Social choice on pollution management: the genossenschaften. Journal of Public Economics 2: 101-146.

Lave, L. B. (1963) The value of better weather information to the raisin industry. Econometrica 31: 151-164.

Lehmann, E. (1959) Testing Statistical Hypotheses. New York: John Wiley & Sons.

Louwes, S. L. (1967) Cost allocation in agricultural surveys. International Statistics Review 35: 264.

Luce, R. D. and Raiffa, H. (1957) Games and Decisions. New York: John Wiley & Sons.

Morgenstern, O. (1973) On the Accuracy of Economic Observations. 2nd Edition, Princeton, N. J. : Princeton University Press.

Moser, C. (1977) The Environment in Which Statistical Offices Will Work in Ten Years Time. CES/SEM.8/3. paper delivered at Statistical Commission and Economic Commission for Europe, Conference of European Statisticians, Washington, D. C., March 21-25, 1977.

Muhsam, H. V. (1956) The utilization of alternative population forecasts in planning. Bulletin of the Research Council of Israel 5C: 133-146.

Nathan, R. P. et al. (1975) Monitoring Revenue Sharing. Washington, D. C.: The Brookings Institution.

National Research Council (1976) Setting Statistical Priorities. Report of the Panel on Methodology for Statistical Priorities of the Committee on National Statistics. National Academy of Sciences, Washington, D. C.

National Research Council (1978) Counting the People in 1980: An Appraisal of Census Plans. Report of the Panel on Decennial Census Plans, Committee on National Statistics. Washington, D. C.: National Academy of Sciences.

National Research Council (1980) Report of the Panel on Small Area Estimates of Population and Income, Committee on National Statistics. Washington, D. C.: National Academy of Sciences. In press.

National Science Foundation (1975) General Revenue Sharing Utilization Project Vol.3: Synthesis of the Formula Research. Research Applied to National Needs, National Science Foundation. Stock No. 038-000-00245-7. Washington, D. C.: U. S. Govt. Printing Office.

Office of Federal Statistical Policy and Standards (1978) Statistical Policy Working Paper I, Report on Statistics for Allocation of Funds. Washington, D. C.: U. S. Department of Commerce.

Office of Revenue Sharing, U. S. Dept. of Treasury (1973) General Revenue Sharing, Final Data Elements, Entitlement Periods 1,2,3. Washington, D. C.: U. S. Govt. Printing Office.

Office of Revenue Sharing, U. S. Dept. of Treasury (1974a) General Revenue Sharing, Final Interstate Data and Allocations, Entitlement Periods 1,2,3. Washington, D. C.: U. S. Govt. Printing Office.

Office of Revenue Sharing, U. S. Dept. of Treasury (1974b) General Revenue Sharing, Final Interstate Data and Allocations, Entitlement Period 4. Washington, D. C.: U. S. Govt. Printing Office.

Office of Revenue Sharing, U. S. Dept. of Treasury (1975) General Revenue Sharing, Interstate Data and Allocations, Entitlement Period 5. Washington, D. C.: U. S. Govt. Printing Office.

Office of Revenue Sharing, U. S. Dept. of Treasury (1976) General Revenue Sharing, Interstate Data and Allocations, Entitlement Period 6. Washington, D. C.: U. S. Govt. Printing Office.

Office of Revenue Sharing, U. S. Dept. of Treasury (1977a) General Revenue Sharing, Data Elements, Entitlement Period 7 and Entitlement Period 8. Washington, D. C.: U. S. Govt. Printing Office.

Office of Revenue Sharing, U. S. Dept. of Treasury (1977b) General Revenue Sharing Ninth Period Allocations. Washington, D. C.: U. S. Govt. Printing Office.

Ono, M. (1972) Preliminary evaluation of 1969 money income data collected in the 1970 Census of Population and Housing. Proceedings of the Social Statistics Section of the American Statistical Association, 390-396.

Pfanzagl, J. (1959) A general theory of measurement: application to utility. Naval Research Logistics Quarterly 6: 283-294.

Raiffa, H. (1968) Decision Analysis: Introductory Lectures on Choices Under Uncertainty. Reading, Mass: Addison-Wesley.

Raiffa, H. and Schlaifer, R. (1972) Applied Statistical Decision Theory. Cambridge, Mass: MIT Press.

Rao, C. R. (1973) Linear Statistical Inference and Its Applications, 2nd Edition. New York: John Wiley & Sons.

Rawls, J. (1958) Justice as fairness. Philosophical Review 67.

Rawls, J. (1971) A Theory of Justice. Cambridge, Mass.: Harvard University Press.

Redfern, P. (1974) The different roles of population censuses and interview surveys, particularly in the U.K. context. International Statistics Review 42: 131-146.

Robinson, J. G. and Siegel, J. S. (1979) Illustrative assessment of census underenumeration and income underreporting on revenue sharing allocations at the local level. To appear in Proceedings of the Social Statistics Section of the American Statistical Association.

Savage, I. R. and Windham, B. (1973) Effects of Bias Removal in Official Use of United States Census Counts. The Florida State University, Department of Statistics, Tallahassee, Florida.

Savage, I. R. (1975) Cost benefit analysis of demographic data. Advances in Applied Probability-Supplement 7: 62-71.

Savage, L. J. (1972) The Foundations of Statistics, 2nd Edition. New York: Dover Publications, Inc.

Sen, A. K. (1970) Collective Choice and Social Welfare. San Francisco: Holden-Day.

Sen, A. K. (1977) On weights and measures: informational constraints in social welfare analysis. Econometrica 45: 1539-1572.

Shryock, H. S., Siegel, J. S., and Associates (1973) The Methods and Materials of Demography, 2 vols., 2nd printing. Washington, D. C.: U. S. Govt. Printing Office.

Siegel, J. S. (1975) Coverage of Population in the 1970 Census and Some Implications for Public Programs, in U. S. Census Current Population Reports, Series P-23, No. 56. Washington, D. C.: U. S. Govt. Printing Office.

Siegel, J. S.; Passell, J. S.; Rives, N. W. Jr.; and Robinson, J. G. (1977) Developmental Estimates of the Coverage of the Population of States in the 1970 Census: Demographic Analysis, U. S. Census Current Population Reports, Series P-23, No. 65. Washington, D. C.: U. S. Govt. Printing Office.

Spencer, B. (1980) Implications of equity and accuracy for undercount adjustment: a decision-theoretic approach. To appear in U. S. Bureau of the Census, Proceedings of the Conference on Census Undercount, Arlington, Va. 1980.

Siegel, J. S. et al. (1977) Developmental Estimates of the Coverage of the Population of States in the 1970 Census: Demographic Analysis, U. S. Census Current Population Reports, Series P-23, No. 65. Washington, D. C.: U. S. Govt. Printing Office.

SRI (1974a; 1974b) see Stanford Research Institute

Stanford Research Institute (1974a; 1974b) General Revenue Sharing Data Study, Vol. III; Vol. IV. Menlo Park, Ca.: Stanford Research Institute.

Strauss, R. P. and Harkins, P. B. (1974) The impact of population undercounts on General Revenus Sharing allocations in New Jersey and Virginia. National Tax Journal XXVII: 617-624.

von Neumann, J. and Morgenstern, O. (1953) Theory of Games and Economic Behavior, 3rd Edition. Princeton, N. J.: Princeton University Press.